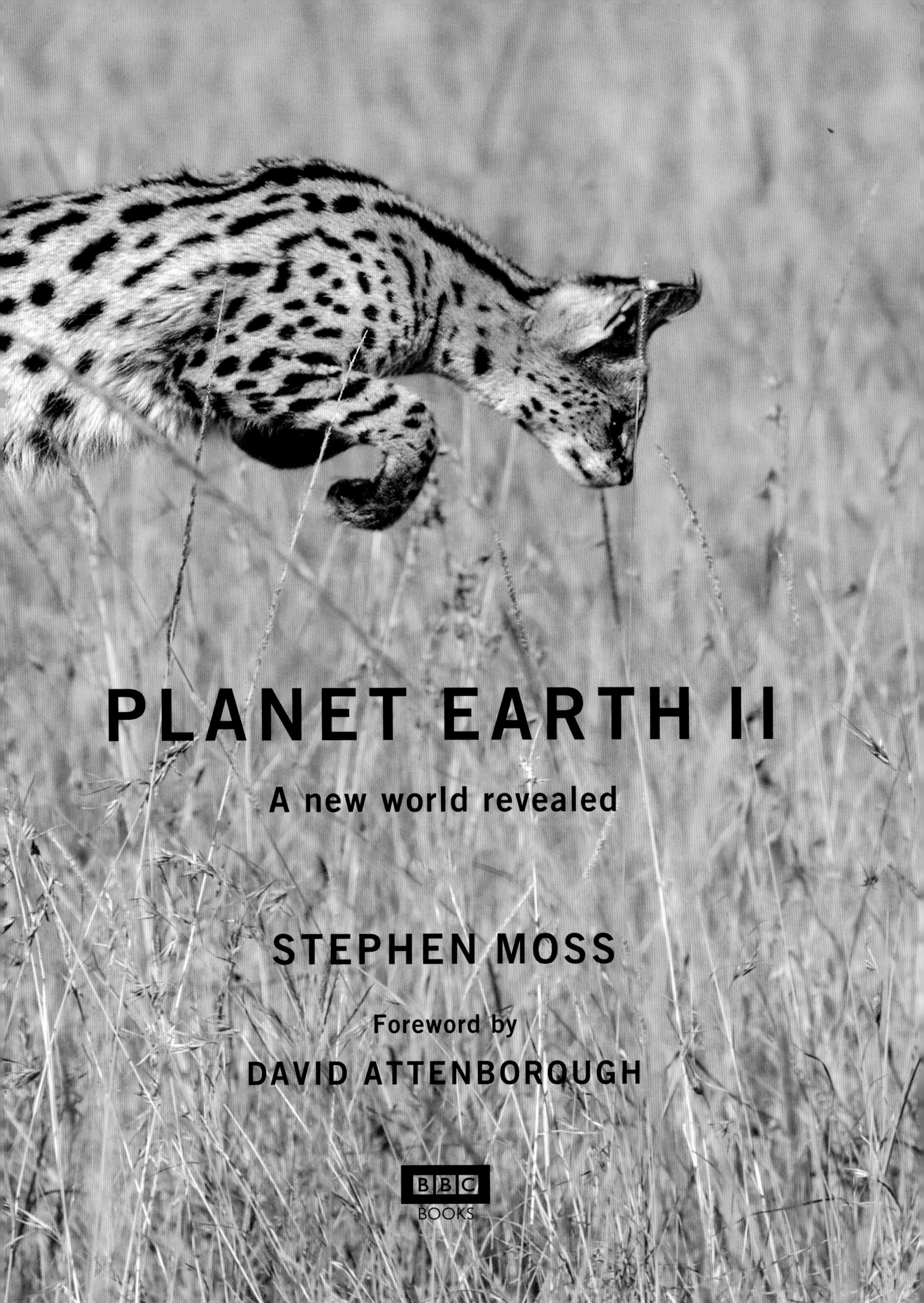

# PLANET EARTH II

A new world revealed

STEPHEN MOSS

Foreword by

DAVID ATTENBOROUGH

BBC BOOKS

# CONTENTS


**Foreword** 6

1 **JUNGLES** 12
Producer: Emma Napper

2 **MOUNTAINS** 50
Producer: Justin Anderson

3 **DESERTS** 84
Producer: Ed Charles

4 **GRASSLANDS** 124
Producer: Chadden Hunter

5 **ISLANDS** 168
Producer: Elizabeth White

6 **CITIES** 212
Producer: Fredi Devas

7 **TALES** 256

**Index** 304

**Acknowledgements** 310

**Picture credits** 312

## FOREWORD
DAVID ATTENBOROUGH

**A cool view of the Earth.** From a hot-air balloon 3,000 metres (10,000 feet) up in the Swiss Alps, David Attenborough surveys the mountain landscape and delivers a commentary for *Planet Earth II,* while being filmed from a helicopter.

Our image of the planet on which we live has changed profoundly over the last few decades. Sixty years ago, almost the only pictures we saw of animals, other than those in our own countryside, were in books. The cinema occasionally showed some, but they were always the familiar ones. Lions, tigers and elephants – maybe. Aardvarks, pangolins and birds-of-paradise – never.

Television began to change that. The electronic cameras of the time were the size of refrigerators and had to be pushed about on wheels. They also needed electric power beyond the capabilities of any battery. So if television wanted to see exotic creatures outside the studio, it would have to use film. But though the movie cameras were theoretically portable when in skilled, strong hands, they were still cumbersome and consumed huge quantities of 35mm film at an alarming rate. They were certainly not the sort of thing you wanted to haul about when trying to creep up on some shy, elusive animal.

But then, in the mid-1950s, things began to change. Film cameramen working for television started to use cameras that until then had been for amateurs only – small ones that took 16mm film.

It was a blissful time for those of us trying to make programmes about natural history. Even in Africa, where many of the animals were so familiar, we could easily find creatures that were quite new to most viewers – gerenuks and porcupines, hornbills and weaver birds, chameleons and turtles. And if we went to other continents, there were all kinds of astonishments – wombats and narwhals, hummingbirds and armadillos, manatees and sloths. Soon all kinds of these exotic creatures began to appear on the flickering box in our living rooms. And with almost every big series, we could promise to show creatures that had never been filmed before.

But electronic cameras were changing. They were getting smaller, more manageable and more reliable. By the mid-1980s, they could be powered by batteries so small and long-lasting that they could be taken to the most remote jungle, the highest mountain and the most desolate wilderness. Some of the cameramen still using the older film cameras, however, resisted these new devices. If something went wrong with a piece of their gear in the back of beyond, they said, you stood a chance of mending it with a screwdriver. You couldn't do that with these new-fangled electronic devices.

But electronic cameras brought huge advantages. Instead of exposing reel after reel of expensive film that could be damaged by humidity or scratches even before it was processed, you could record everything on tiny cards. You could use them with extravagance because memory cards were both wipeable and cheap. And instead of having to wait – sometimes for months – for the film to come back from processing, you could replay your pictures immediately to see if you had actually recorded what you wanted.

Those advantages were, in themselves, enough to persuade most of us to make the change, and within a decade, almost all natural history film-makers had done so. It was then that we began to realize that we had, almost unknowingly, entered a new era. Most importantly, electronic pictures could be viewed on screens separate from the camera that was producing them. You could put a miniature camera, unnoticeably small, beside a bird's nest and watch what was happening while relaxing in a tent a mile away. No longer did you have to sit with your finger on the start button, ready to press it when – or preferably, just before – an animal did something interesting. Now you could use an instant-replay device that automatically recorded the 30 seconds before you pressed that button. We could even arrange for an unattended camera to be turned on by an animal's movement or turned off by the

▲ **Last wilderness spectacle.** The Porcupine caribou herd moves across Alaska's Arctic National Wildlife Refuge, from its calving grounds to foothills and mountains. It's one of the last great migration spectacles, filmed for *Planet Earth II*. The calving grounds are now threatened by oil development.

protracted absence of it. We called these last devices 'camera traps' even though they trapped nothing except images. None of these things had been possible with film cameras.

For those of us making natural history films, it was liberation. For most of those watching, however, the change barely registered – except that, year after year, pictures came from increasingly remote places and showed events, animals and behaviours that no one had ever known about before. Memorable series followed, one after the other – *Planet Earth*,

*Blue Planet*, *Frozen Planet*, *Africa*, *The Hunt* – each with new wonders and revelations.

And now comes *Planet Earth II*. I remember 20 years ago, when planning a new series, considering a sequence about the snow leopard. No one had ever filmed this marvellous and then still-mysterious animal in the wild. We researched it in great detail, but the more we discovered the more improbable it seemed that we could ever film it. Eventually we decided that the animal's rarity and the vastness and emptiness of the areas in the Himalayas where it lived made the project impractical and we abandoned it.

Ten years ago, the team producing the first *Planet Earth* series were bolder. Two highly experienced natural history cameramen, as is related later in this book, took on the job. They worked patiently for more than two years. In the end one of them secured a memorable shot of a snow leopard stalking, chasing and finally pouncing on a Himalayan

goat – which then, with astonishing strength, shook off its attacker and escaped by leaping into a river. For a picture of a hunting snow leopard it was surely unsurpassable. But now new technology has allowed *Planet Earth II* to go one step further. The producers decided to try to film the social life of this most solitary of all the big cats. To do that, they deployed 20 camera traps. The results are not only thrilling for television viewers, they also revealed aspects of snow leopard behaviour that were new even to the scientists who have been studying these magnificent animals for years. And this is only one of a whole range of astonishments in this new series.

Does all this matter? Has all this technical invention and human endeavour provided us with anything more than a wonderfully watchable and visually unforgettable television series? I believe it has. Since those early days, the human population of this planet has tripled. As a consequence, the space for the other creatures with which we share the world has become more and more restricted. Series such as *Planet Earth II* manage to bring us greater and greater understanding of the natural world, the way it works and what it needs if it is to continue to survive.

And its survival could not be more important to us. We depend on the natural world for all the food we eat, for the very air we breathe. Its health is our health. Its survival is essential for ours.

▲ **Dangerous encounter.** The female snow leopard star of the series (right) attempts to drive away a male who could threaten her adolescent cub – behaviour never-before observed, let alone filmed.

# 1
# JUNGLES

**RAINFORESTS ARE THE MOST** biodiverse places on our planet. Of the latest estimate of roughly 8.7 million species of plants and animals on Earth, more than half live in tropical rainforests; 25 hectares (about 62 acres) of Amazon rainforest contains more than a thousand species of trees, compared with fewer than a hundred in the whole of the British Isles. Rainforests are also the lungs of the planet, turning carbon dioxide into oxygen, which is essential to sustain virtually all life.

The Amazon region is home to well over 2 million species of insect, 40,000 different plants, more than 400 mammals and 1300 birds – that's more species of bird than in the whole of North America and Europe combined.

Rainforests such as this were the original home of many of the foods that we take for granted, from fruits and vegetables to spices and nuts. And scientists are still discovering new drugs from rainforest plants. With such an abundance of things to eat and places to live, it is hardly surprising that more species have evolved here more rapidly and more often than anywhere else.

These vast labyrinths of trees are not only the 'cradle of life' but also the 'museum of life', with species spreading out to colonize other habitats. This is, after all, where the ancestral primate that gave rise to humans evolved, before it ventured out onto the grassy plains of Africa. But in this paradise, there is a major drawback. Having everything you need does not, paradoxically, mean that life is easy. With such a multiplicity of species having evolved there, rainforests are the most competitive of places – just finding your own space can be a challenge – with as many dangers as opportunities. To survive and thrive in this crowded place, you need to find your niche and adapt accordingly. That is the challenge faced by every creature that lives in a jungle.

(previous page) **Jungle dawn.** Sunrise over the Borneo rainforest in the Danum Valley, Sabah, Malaysia – home to a huge diversity of plants and animals, including orangutans.

◄ **Frog gathering spot.** Misfit leaf frogs mating beside a rainforest pond in La Selva, Costa Rica. Such mass mating is usually triggered by a huge rainstorm at the end of the dry season.

## The five-limbed swingers

▶ **Hanging out.** A black-handed spider monkey hangs from the branches of a rainforest tree in Costa Rica, using her prehensile tail as a fifth limb, while her baby holds onto her with its tail. She will feed mainly in the upper part of the canopy, swinging from one slender branch to another in the search for ripe fruit – a spider monkey's preferred food. Long limbs and a long tail are essential for such treetop living.

From high above, the forest canopy appears to be a uniform blanket of green. But this is very far from the truth. Rainforests are a mosaic of different species of tree, some with tasty leaves and fruit, others inedible and some even poisonous. Nor is each kind of tree confined to a specific area – individual trees of the same species may be miles apart. So if you depend on one particular species for food, you need to know how and where to find it. Spider monkeys have learned to do just that, by making a mental map of their home range.

Spider monkeys are found in the tropical forests of South and Central America, from Brazil in the south to Mexico in the north. Light and agile, they live in the rainforest canopy where they feed mainly on a high-energy diet of ripe fruit, which they find using their excellent colour vision and forward-facing eyes. Their name comes from their extraordinarily prehensile tail, which can act as a fifth limb, enabling them to move around the upper layers of the trees with practised ease as they go in search of fruit. Sociable creatures, they live in troops of 30 or more, though they may split into smaller groups to search for food.

Adult spider monkeys are excellent climbers and foragers. Their babies, like other young mammals, learn by exploring their surroundings and discovering through trial and error the best places to find food and then the best way to get hold of it – sometimes the most difficult part.

In this complex, three-dimensional world, so far above the ground, a mistake can prove fatal. Following its mother and the other animals in the group, a youngster soon discovers that not all trees are the same: some are smooth and slippery, others too far apart to leap between. So long as it keeps at least two of its five 'limbs' (including that amazing tail) in contact with the tree, it is fine. Fortunately, if it slips, which it will do as it becomes more confident, it has the equivalent of a safety harness to save its life. When a spider monkey falls, its tail automatically locks around the branch to which it is attached. So though the baby is left hanging in mid-air, it is safe, and a parent can rescue it, using its body to bridge any gap. Prehensile tails appear to have evolved independently in a number of New World high-living monkeys but are most advanced in spider monkeys, used for both travelling and feeding – especially when they need to get to fruit near the end of a long, thin branch.

▲ **Flash warning.** A male draco lizard guards his territory on a rainforest tree in Malaysia, opening his dewlap flag to warn off another lizard. The dewlap is also a useful device for signalling his availability to females.

◄ **Gliding technique.** Expanding his ribs like umbrella struts so the skin between them stretches to form wings, a male draco lizard glides to a different perch, using his long tail as a rudder. His neck flaps also help in the controlled glide.

## The flyaway lizard

Mastering the forest freeways can be even harder when you are small. For a draco lizard in the jungles of Southeast Asia, walking on the forest floor may be dangerous, but a treetop life has its challenges. What to a monkey would be a simple leap is for a lizard a chasm. And it faces an even more pressing problem. Like many small animals, it has a rapid metabolism, and it requires a constant supply of energy in the form of ants or termites.

So it has become an expert in finding ant trails, scent-marked by the ants as they travel from their nests in search of food. All the lizard has to do is wait for an ant convoy to arrive – as long as another draco lizard is not already in position on the tree guarding a territory. If the rival is larger, it will flash open a flap under its throat as a warning. The smaller reptile has no alternative but to leave, but it can do so without having to climb. It extends its extra-long ribs to form wings, launches itself into the air and simply glides to a neighbouring tree, steering with its long tail, to start the search for food once more.

## Birds shaped by flowers

When it comes to fight or flight, one group of tropical birds has it covered. Hummingbirds are among the most specialized yet also one of the most successful of all the world's bird families. There are well over 300 species, found throughout North and South America. They range from the smallest bird in the world, the bee hummingbird of Cuba, to the thrush-sized giant hummingbird of western South America. But all have one thing in common: they have evolved to exploit a huge range of sources of nectar, which they reach using their rapid and incredibly manoeuvrable flight.

Hummingbirds are a good example of how complex habitats such as rainforests support a very wide range of species from the same family, each of which has evolved to specialize in a subtly different niche, exploiting the abundant resources in a range of ways. In the case of hummingbirds, they feed on nectar at different levels (some high, some low), at different times of day or on different flowers. Some even have a difference in beak length between the sexes, so that the males and females are able to exploit different feeding opportunities in the same home range. With such a high-energy lifestyle, they need to feed often and regularly, and for their size, they are feisty and may fight with one another to gain access to the flowers that contain the most nectar.

With so many species of hummingbird competing for the same food, it pays to be different. And they do not come much more different than the sword-billed hummingbird, found about 2500 to 3000 metres (8200 to 9850 feet) above sea level in the humid montane forests of the Andes. As its name suggests, the sword-billed hummingbird's beak is long – very long. Indeed at 9–11cm (3.5–4.3 inches), it is even longer than its body – and proportionately longer than any bird's bill in the world. It has evolved to exploit flowers that have very long corollas, such as particular species of passionflower – too long for shorter-billed hummingbirds to reach the nectar inside. This way the sword-billed hummingbird avoids competition and finds its own niche in a crowded world, and the flowers have their special pollinator.

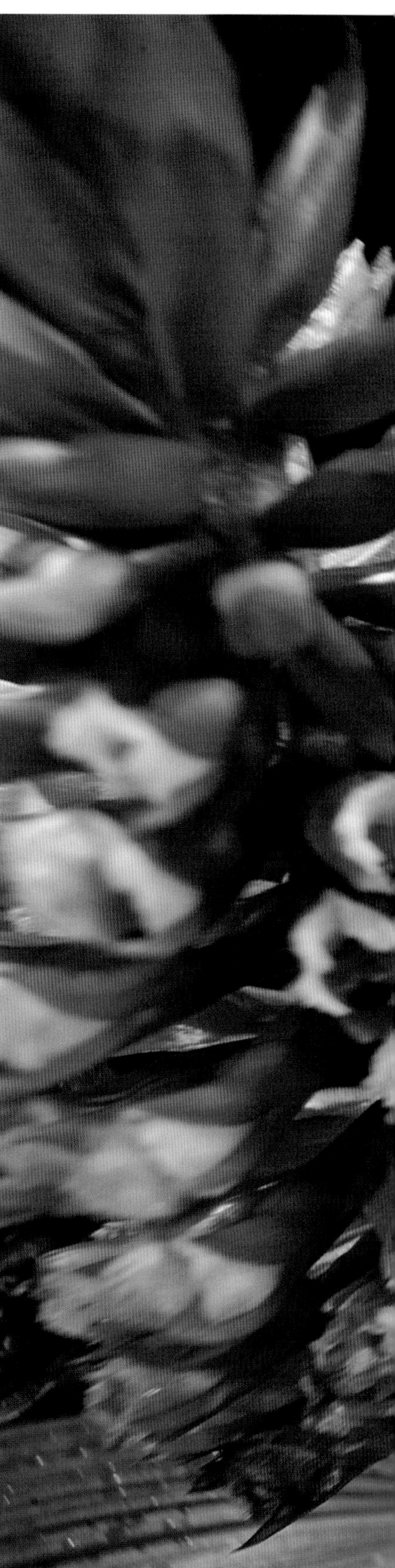

▶ **Nectar pollen swap.** A female volcano hummingbird sips nectar from a cloud-forest *Eleanthus* orchid and in the process gathers pollen on her beak. She makes a regular circuit of the same flowers, and when she has young, may visit many hundreds in a day.

◄ **Flower fight.** Chestnut-breasted coronet hummingbirds fight over access to flowers in mountain forest, Ecuador. The male clinging to the flower is probably the territory holder, defending his nectar source, which has florets at the ideal length for his relatively short bill.

► **Long drink.** A sword-billed hummingbird feeds at a passion flower in the Ecuadorian cloud forest. Its 10cm (4-inch) long bill matches the extremely long corolla of the flower, and it is the only significant pollinator of this High-Andean species. It will visit this plant on its regular circuit, checking which new flowers are open for nectar gathering.

But the sword-billed hummingbird also faces two problems caused by that incredibly long bill. At times when nectar is scarce, hummingbirds catch and eat tiny insects, which may also be vital protein for feeding to their chicks. For the sword-billed hummingbird, managing to grab a flying insect with its huge appendage can be tricky. It overcomes this by opening its bill wide while hawking in flight like a swift, maximizing its chances. The other problem posed is how to use that enormous bill to preen its feathers. The answer is simple – to use its feet.

## Stinkbirds and enchanted dolphins

All rainforests have regular and heavy falls of rain. Though life-giving, this also poses both immediate and long-term problems for those creatures that live there. When the rain begins to fall – and in some places it does at the same time every day – many creatures seek shelter. The smaller ones such as insects and songbirds hide in holes and crevices in tree trunks and branches, and the larger ones sit it out under whatever shelter they can find. When it rains hard, orangutans may even use large leaves as makeshift umbrellas.

Regular heavy rainfall results in the creation of specialist habitats. One such lies along the huge Araguaia River in central Brazil, where the forest is flooded for a substantial part of the year. This kind of seasonally inundated rainforest is known as várzea, and occurs throughout the Amazon Basin. In

▲ **Daily downpour.** A regular afternoon rainfall in tropical rainforest, Australia. With abundant fresh water and sunlight all year round, tropical rainforests – now covering just 6 per cent of the land – provide perfect growing conditions and contain more than half of all the planet's known plants and animals.

▶ **Flooded forest.** Sunlight filters through the tree canopy in the flooded forest along the banks of the Amazon River in northern Brazil, home to many specialist animals, including river dolphins.

◀ **Clawing back up.** A hoatzin chick climbs in the vegetation near its nest, using its 'hand' claws as well as its feet. The nest is over the river, and the chick can avoid a predator by deliberately falling into the water. Once the danger has passed, the chick uses its wing and feet claws to clamber back into its nest. The hoatzin is the only living bird to climb using claws on its wings, which are lost by the time the bird becomes an adult.

▶ **Mother's watchful eye.** A female hoatzin keeps watch on her chicks. She is aided by a team of helpers, usually youngsters from the previous year that are not yet mature enough to breed, an example of cooperative nesting rare in the bird world.

the rainy season, very high rainfall causes water levels to rise – sometimes as much as 10–15 metres (33–49 feet). The creatures that live here are adapted to this watery world, and they include one very peculiar bird, the hoatzin, also found in tropical-forest wetlands elsewhere in northern South America. This species is so different that not only does it have its own family but also its own order: the Opisthocomiformes – a name that refers to the bird's shaggy crest and means 'wearing long hair behind'. The fermentation in its digestion also causes it to smell, and it is known locally as stinkbird.

The pheasant-sized hoatzin has a staring red eye that gives it a rather surprised expression. But what makes it unique among birds is the fact that the chicks have claws on their wings, reminiscent of birds such as *Archaeopteryx*, which lived millions of years ago in the era of the dinosaurs. If threatened by an intruder, a baby hoatzin will jump out of its nest into the water and swim – even under water if necessary – using its feet and wings. Once danger has passed, it uses the claws on its wings to clamber back into the nest.

As an adult, a hoatzin has lost the claws on its wings and is unable to swim. It is not even very good in the air. That's because its specialized diet of leaves means it has a huge foregut to digest them, comprising as much as

25 per cent of its overall body weight when full of food. To make room for this, its sternum (or breast-bone) is much smaller and lighter than it should be for this size of bird. This means it is only just able – with considerable effort – to travel even short distances on the wing. It is also unable to walk easily and must spend a lot of its time digesting its gutful of leaves, making the hoatzin one of the most sedentary of birds.

The largest creatures to be found in the flooded forest are the three river dolphins: the Amazon river dolphin, or boto, of the Amazon and Orinoco river basins; the Bolivian river dolphin, found in the upper Madeira River basin (the Amazon's largest tributary), isolated from the Amazon River by massive rapids; and the Araguaia river dolphin – a species that is so rare and elusive that it was only described for science in 2014. The Araguaia river dolphin probably became isolated and evolved into a distinct species 2 million years ago, when the Araguaia–Tocantins river basin in Brazil became separated from the rest of the Amazon river system by waterfalls and rapids.

These three South American species are the largest of the world's river dolphins, reaching up to 2.5 metres (more than 8 feet) long and weighing more than 200kg (440 pounds). They have more flexible bodies and softer skin than marine dolphins, making them less rapid swimmers but far more manoeuvrable, especially from side to side, a crucial advantage when chasing prey in labyrinthine river systems. And they can paddle forwards with one large flipper and backwards with the other when twisting and turning in the flooded forest. They have long snouts lined with sharp teeth, the front ones for grasping and the rear for crushing their prey, which includes catfish and crabs.

As large river mammals, the dolphins inevitably have a close relationship with the people who live alongside the rivers. This can be uneasy: the dolphins get struck by boats and injured or caught up in fishing nets and drowned. And in the case of botos, settlers moving into the area catch them for use as fish bait. But botos are also regarded by the forest peoples as *encantados* – enchanted creatures that can transform from animal to human and back again.

► **Forest dolphin.** An Amazon river dolphin, or boto, swims through the flooded forest in the muddy, tannin-stained water of Rio Negro, Brazil. It has little need of vision, using echolocation to hunt and navigate through the maze of tree roots.

## The biggest surprise in the jungle

The newly discovered Araguaia river dolphin is, like its close relatives, a pale pinkish-grey, with a domed head and tiny eyes (when hunting in the murky river water, good sight is not much of an asset). The main physical difference is in its slightly wider skull, though it also has fewer teeth. The population probably numbers less than a thousand in a river system fragmented by dams. Very little is known about its behaviour, but observations made by the BBC team when filming these shy creatures revealed that they do come together in pods and even hunt in teams.

**1 Blowholes in the forest.** In the flooded forest of the Araguaia River, the domed heads of two river dolphins appear – their blowholes (breathing holes) just visible – as they navigate the shallows.

**2 Boat view.** Side-on, a dolphin reveals its small dorsal fin and bulbous forehead, the melon. The size of the melon is linked to the vital importance of echolocation for hunting and communicating in such murky water.

**3 Aerial view.** A group of the dolphins feeds near the surface, possibly hunting cooperatively. With such murky water, an aerial view from a drone is the only way to see how many dolphins might be feeding or socializing.

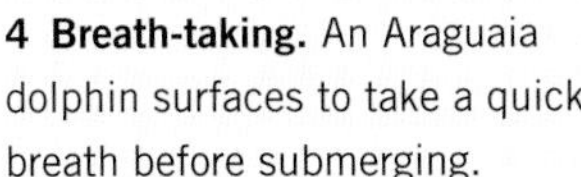

**4 Breath-taking.** An Araguaia dolphin surfaces to take a quick breath before submerging.

## The predators' predator

The richness of the riverine ecosystem and variety of species in Amazonia, Brazil, means there are also many predators. Dazzling kingfishers and a wide range of herons and egrets hunt for fish, while caimans, giant otters and predatory fish such as dorado seek out their prey in the dark, murky waters. But all yield to the ultimate predator, the jaguar. This fearsome cat regularly patrols the edge of riverbanks in search of anything it can catch.

The jaguar is an excellent swimmer and, like most big cats, an opportunist, hunting a wide range of prey, from freshwater turtles and armadillos to capybaras and deer. What it uses is stealth. Sometimes it lies in wait, ambushing any animal that comes to drink. At other times it stalks its prey to within pouncing distance and then kills with a bite to the throat or the back of the neck.

The smaller female jaguar generally hunts smaller prey, though that still includes capybaras – large enough to make a satisfying meal. She may hide in the curtain of trees along the riverbank, camouflaged by her blotchy coat. But however careful a female tries to be, if she is accompanied by a cub, she is often frustrated by her cub blowing her cover. But if it manages to stay quiet, she will stalk a capybara or even a small caiman (a crocodilian) until close enough to leap into the water and kill it, at the same time giving her cub a hunting lesson.

In the Amazon rainforest, there is so much food that even top predators such as the jaguar can live at relatively high densities, though there is still competition, especially between big males. A male can weigh up to 120kg (264 pounds), but only the oldest, the most experienced and the most daring will try to kill an even larger predator, the caiman. To avoid injury, the jaguar grabs the caiman at the only weak spot in its scaly armour, the back of its skull. He then hangs on until the writhing animal finally gives up the fight and he can let go. Jaguars that hunt caiman often have battle scars, but the reward is great – a huge reptile is a feast.

◀ **River watch.** A female jaguar hunts along a riverbank in Brazil. Jaguars frequently use riverbanks for stalking capybaras (huge rodents) and other prey, in and out of the water.

▶ (next page) **Caught napping.** A jaguar drags a yacare caiman into the shallows in Brazil. These crocodilians are small enough to be taken by jaguars. In flooded forests, where there are few large mammals, reptiles form a minor but important part of a jaguar's diet.

## Hiding in plain sight

Being small makes it easy to hide, but it also means many bigger creatures that could eat you. The various species of leaf-tailed geckos – found only on the island of Madagascar – are tropical-forest specialists, named for their broad, flat tails, which really do look like leaves. This helps them hide, as does their skin, which can resemble the bark of trees or even dead leaves. They spend much of the day hiding in plain sight, perched motionless while basking in the sun, waiting for night when they will stalk their own food – insects and other invertebrates.

But one group of amphibians matches the leaf-tailed geckos in their ability to stay hidden, even when in full sight. Like the geckos, glass frogs merge with the plant on which they sit. But they do this by being see-through. More than 120 different glass frogs are found in the tropical and equatorial regions of Central and South America. Some are tiny – not much bigger than the top of a human thumb – and though basically lime-green, the skin on the abdomen of many species is translucent, revealing some of their

▲ **Dead-leaf pose.** A satanic leaf-tailed gecko hides in the day using its remarkable camouflage. Found only on Madagascar, this tiny lizard can cling on while moving from leaf to leaf. Like all geckos, the under-surface of its bulbous toes are covered with hundreds of microscopic hairs that effectively make them sticky.

► **Lichen-bark pose.** A Madagascan mossy leaf-tailed gecko hides in the day, head down, flat against a tree trunk, using camouflage that matches the lichen and moss on the bark.

internal organs. They live mainly in the foliage of trees, especially ones alongside streams and rivers, and have expanded tips to their digits that allow them to hold onto leaves.

It is the male glass frog that raises the young. He first finds the ideal spot: a smooth leaf facing away from the sun, protected from the elements, out of sight of birds and hanging several metres above a flowing stream, into which the tiny tadpoles will drop when they eventually hatch. There

▲ **The leaf nursery.** On the underside of a leaf in the Ecuadorian rainforest, a tiny male Atrato glass frog guards two clutches of eggs from two different females. Eventually, the tadpoles will wriggle out of the gelatinous egg mass and drop into the stream below.

he waits, calling at night to attract a female to his leaf territory, fighting off any males wanting to take it over. The female merely lays the eggs on the leaf and departs. The male then takes care of them until they hatch, which can take three weeks, depending on the species. Though he can't fight off predators such as snakes, he can dissuade insects, such as wasps and also parasitic flies that try to lay their eggs on the spawn (the resulting maggots would eat the frog embryos).

## The ninja guard and the hunter wasp

Glass frog eggs are laid on the underside of leaves and are hard to see, as is the father who guards them, but wasps can locate them and will prey on eggs and tadpoles. If the developing tadpoles feel the vibrations of a wasp trying to penetrate the jelly, they hatch prematurely and drop into the water below. But younger egg batches are more vulnerable, so the male positions himself beside them. Being vulnerable to predators himself, he keeps flattened and still, legs folded in, almost resembling an egg mass, but if a wasp lands nearby, he gives it a ninja-style kick and keeps kicking until the wasp goes in search of easier prey.

**1 Father on his leaf.** A male reticulated glass frog on his leaf in a Costa Rican rainforest. On the underside of the leaf are developing broods, each laid by a different female. The leaf nursery is over a stream, into which the developed tadpoles will eventually drop.

**2 Predator alert.** The male frog watches as a wasp attacks one of his broods. If he jumps across to drive off the wasp, he risks being spotted by daytime predators.

**3 Tadpole grab.** The wasp penetrates the sticky jelly, but though it may eat one tadpole, the others are already wriggling out of their 'jelly ponds' for a premature birth and descent into the stream below.

**4 Kick-boxing.** The male kicks out at the wasp. Strangely the wasp is trying to land on the frog. One theory is that the father's reticulated pattern makes him into a decoy, attracting the wasp to what it thinks is an egg mass, when all it will get is a kick.

▲ **Lurking spider.** A fishing spider waits beside the glowing fungus, appearing to have learned that insects are attracted to the luminescence and will provide easy prey.

◄ (top) **Night glow.** A click beetle lands on a glowing fungus, lured there at night by the expectation of copulation. But the nearest it will come to sex is getting covered in the spores of the fungus, which it will then take to the next fungal lure it visits.

◄ (bottom) **Luminous lure.** A confused click beetle crawls over the gills of the glowing fungus – the world's most luminous – flashing its two lights, presumably in anticipation of communicating with a female. Female click beetles, by comparison, will sit and glow constantly to attract males to them.

## The decoy glow

The edges of the jungle are full of life, but its heart can be so dense and dark that organisms need to evolve special features to thrive there. Some fungi make use of the darkness. For a few weeks every year, they send up their spore-producing fruiting bodies – toadstools – which glow in the dark like beacons. These attract insects, in particular, luminous click beetles (they have two headlight-like spots on the thorax), which presumably mistake the fungus from a distance as the lights of potential mates and land on them. Once they realize their mistake, the beetles fly off, but not before they have become covered with spores, which they then distribute far and wide.

The fungi are not the only organisms to take advantage of the attraction of this spooky light. One type of spider appears to have learned to hang around the fungi, waiting for unwary beetles to land there. When a beetle makes this fatal mistake, the spider pounces and grabs an easy meal.

## How plenty inspires beauty

Even in the densest rainforests, trees do occasionally fall, allowing small pools of light to reach the forest floor. In Papua New Guinea, these are the spots where one of the most bizarre and colourful birds in the world chooses to display.

The Wilson's bird-of-paradise may be small – about the size of a song thrush – but he makes up in energy what he lacks in size. Before he begins his performance, he clears the arena of leaves and any other extraneous material, and also strips the leaves from the sapling on which he will later perch to perform. As dawn breaks, he begins to call to attract females. Using a series of penetrating sounds, together with clicks and whistles, he culminates with a loud whip-crack.

His ritual display starts as the sun rises, illuminating his tiny territory through the hole in the canopy. If a female arrives, he responds by perching at the bottom of the bare sapling and adopting a frozen posture. Then he dances, leaning to and fro, showing off the amazingly bright colours on his unfeathered crown (pale blue), upper mantle (yellow) and back (orange). Because he stays low and the drabber female looks down on him from

► **Courtship by colour.** A male Wilson's bird-of-paradise in characteristic 'frozen' posture on the stem of a sapling, about to perform to the female above him. His chosen forest-floor arena is well lit, and looking down from above, the female will see the full glory of his colours, his breast feathers shining brilliant green and the inside of his mouth flashing a brilliant yellow as he calls.

▼ **Clearing the arena.** The male Wilson's bird-of-paradise prepares his arena for the arrival of the female, clearing away the leaves around his performance sapling.

above, the light from above helps him stand out. He brings the display to a close by flashing his bright colours at her while waving the wires of his tail.

Meanwhile, at daybreak, at the top of the canopy, another species is also courting. A red bird-of-paradise – resplendent in his green, yellow and chestnut-red costume, with red flank plumes and long, ribbon-like tail feathers – displays to watching females, attracted by his loud, nasal calls. But he is not alone: up to ten males may gather in the highest treetops to dance in the early morning and late afternoon, choosing defoliated branches on which to perform. The females then choose from the performers.

Both the Wilson's and the red birds-of-paradise live on the same tiny island off the west coast of Papua New Guinea; both are closely related, yet both behave completely differently. One is a showy extrovert, competing together with his fellow males at the top of the forest canopy; the other prefers a dance-pole and stage on the forest floor and to reveal his skills and beauty on his own patch.

The reason birds-of-paradise can spend so much time and effort on elaborate and complex courtship displays is that these rainforests offer plentiful food all year round. So they do not have to spend their precious time and energy constantly searching for something to eat – as, say, birds living in more hostile habitats such as deserts or mountains have to do. But though the females may be drab compared with the males, they are the ones driving this perfection of plumage and display through the mechanism of sexual selection. They do the choosing, and while the males stay in one place to perform, the females roam around the forest looking for the best males – always knowing that there will be food available wherever they go.

When they have mated, the males then have nothing to do with incubating the eggs or bringing up the young. Again, because there is such a superabundance of food in this rich, complex ecosystem, the females are able to raise a family on their own without the need of a male to help feed the chicks.

▶ **Treetop performance.** On his treetop display post, a male red bird-of-paradise displays to a female, who seems impressed by his performance. He has flipped upside-down, spreading out his plumes, his tail wires framing his wings in a heart shape. If she stays watching, he may well dance along the branch, flipping his body and tail from side to side. But if she doesn't give him full marks, she will just fly off.

## The cousins we left behind

The abundance of food all year round makes living in rainforests in many ways easy. But the extraordinary range of species creates the pressure of competition, which in turn causes animals to evolve further specializations that create unique lifestyles. On the island of Madagascar, millions of years of isolation from other landmasses have allowed a degree of unrivalled specialization among the primates – the lemurs – which until the arrival of humans, shared the forest with few other mammals. They evolved into more than 40 different species (see page 184), ranging in size from a mouse to a gorilla, exploiting different food and places to live. Today, the largest lemur still in existence is the spectacular indri.

Indris live in family groups, with a male and female and their offspring of different ages. They are among the most tree-bound of the lemurs, moving around by climbing and leaping between trunks and branches, but they only have a very short tail, unlike the other large lemurs. They proclaim their presence and warn off other groups of indri using loud songs that penetrate far through the forest. They need their territorial space – indris eat mainly leaves, with a preference for young leaves, and so have to feed regularly to get enough sustenance, needing large areas of forest for this. In fact, 40 per cent of their active time is spent feeding and most of the rest of the time spent digesting. But they tolerate other species of lemur that are not direct competitors for the same food, and many different lemurs live together in the same eastern forest of Madagascar.

The local Malagasy people regard them as cousins, telling a story of how in the past humans chose to leave the forest and build vast cities, while the indris chose to stay behind in the forest. Indeed, African jungles were our ancestral home. The primates that gave rise to our species evolved in this complex, crowded place, and like other successful forest creatures, coped by finding their special place among the other millions of species living there. And though our species evolved many millions of years ago, when our ancestral primate headed out onto the vast grassy plains of Africa, we still retain the ingenuity, vision, handiness and family bonds that all forest primates have.

◄ **Cousin indri.** An indri feeds on shoots and leaves in the forest canopy in Madagascar. It is the largest lemur in existence and the one most resembling higher primates such as humans.

# 2
# MOUNTAINS

**THE COMBINATION OF EXTREME** weather, rugged terrain, precipitous slopes and a lack of food and water to be found on mountains present both plants and animals with some of the toughest physical challenges on the planet. Yet for those creatures that do manage to survive in such inhospitable places, there are benefits – not least the lack of competition from rivals that can make life so difficult in the more biodiverse places such as jungles and grasslands.

The key characteristic of mountains from an ecological point of view is the rapid and major change in climate and vegetation that occurs with increased altitude. It is often said that climbing Mount Kilimanjaro, which at almost 5900 metres (19,500 feet) is the highest peak in Africa, is akin to experiencing four seasons in four days. This does not just refer to the dramatic and often sudden changes in the weather but also to the extraordinary variety of habitats, from dense equatorial forest at the base, through open heathland on the lower slopes, dry desert-like conditions on the upper slopes, and snow and ice fields towards the summit.

Mountains also create ecological isolation. The term 'sky islands' was originally coined in the 1940s to refer to the high peaks of southeastern Arizona, in the USA. Here many of the plants and animals on each mountain were isolated from the surrounding lowland desert and also from one another, and some had evolved into different species, found only on a particular mountain. Others had become 'relict species', refugees from a time when the climate would have been far cooler and

(previous page) **Andes ascent.** Chilean flamingos fly past the glacier-covered peaks of the Torres del Paine National Park, Chile. They are one of three species of flamingos that have adapted to live at altitudes of up to 4500 metres (almost 15,000 feet) above sea level on the high plateau of South America's central Andes.

◀ **Islands in the sky.** Mount Kenya – an ancient extinct volcano and the highest mountain in Kenya. It rises through a gradation of habitats – mountain forest, alpine heath and cushion vegetation, including these slow-growing giant groundsel plants, near the icy summit.

their ranges covered a much larger area. Ecologically, scientists have found that these mountains are similar to oceanic islands surrounded by sea.

The nature of weather and climate is different on mountains too. As altitude increases, so the air becomes cooler, the wind stronger and the weather systems often more unpredictable. All these factors combine to make living in the mountains challenging, especially if it is all year round.

Our image of a montane animal is of a mysterious loner such as the snow leopard, which roams Asia's great Himalayan range, or the mountain lion of the South American Andes. Yet even these solitary creatures can be more sociable and gregarious than we might expect. Moreover, other montane animals do live in large, social groups, enabling them to work collectively to combat the problems they face in this harsh environment. These include mountain goats, able to climb steep slopes with ease; and on the high tops, there are some truly surprising colonies, such as the congregations of flamingos on lakes in the High Andes.

## Masters of the mountain air

For any creature that makes its life in a high mountainous region, the main challenge is finding food, which in these comparatively bare places can be scarce. So a bird such as a golden eagle has to cover a vast territory to find it. Golden eagles are found right across the Holarctic region of the northern hemisphere: in much of North America as far west as Alaska and from the Scottish Highlands all the way east to Japan and south to the Atlas Mountains of North Africa. In many of these mountainous regions, golden eagles are the top predator. Like many raptors, they have acute eyesight, enabling them to spot their prey from a great height as they soar overhead or an easy meal in the form of a dead animal. They also have binocular vision, another major advantage when searching for and catching their prey.

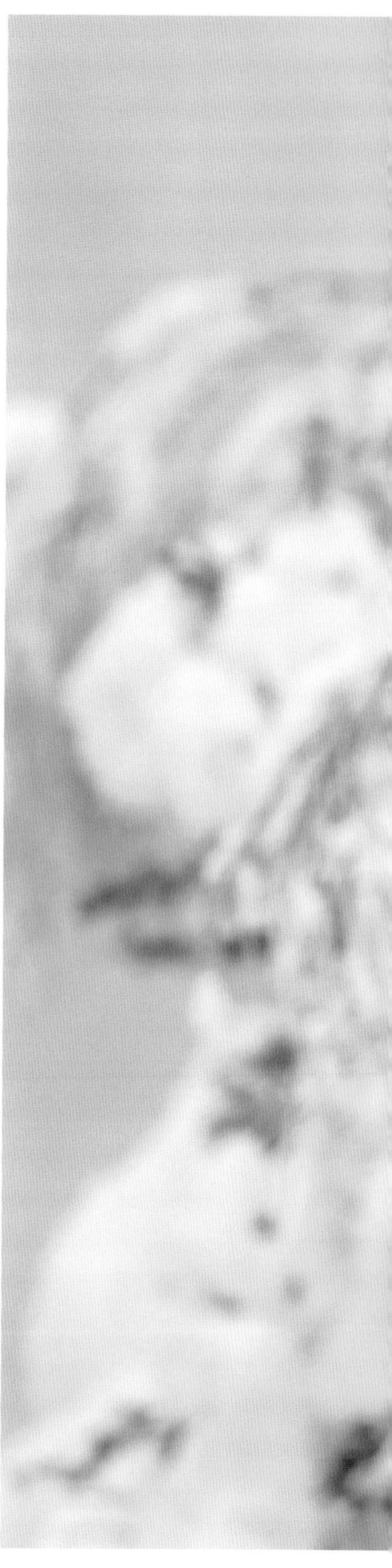

► **Powering down.** In the mountains of Norway, a golden eagle descends to feed on carrion, displaying its huge wingspan – up to 2.2 metres (7.3 feet). The tips of its primary feathers curl up as it manoeuvres into a landing position. Not only can its powerful talons tear the flesh of a carcass, they can also crush its bones.

Once an eagle has targeted its victim – perhaps a ptarmigan, a mountain hare or even a baby goat – it pursues it relentlessly, using its broad wings to gain speed, manoeuvring with surprising skill for such a large bird, before plunging its sharp talons into the flesh. Having killed an animal, the eagle rips it apart using its powerful beak and talons and feeds until it can eat no more. After all, it can never be sure where the next meal will come from.

In winter, however, many eagles use a very different tactic. Prolonged spells of cold, icy weather and heavy snowfalls make life tough for most creatures. But this is to an eagle's advantage. They search for carcasses of animals that have died of starvation. Perching on a crag, an eagle will watch for the noisy activity of other scavengers such as ravens and crows, which often leads it to a convenient meal.

It is crucial that golden eagles get enough food during the winter, for they are one of the earliest of all mountain birds to begin breeding. They usually pair up before the start of the year and return to their nest site in January. The nest is huge, built out of sticks either in a tall tree or on a narrow cliff ledge. If the eagle's prey is mainly heavier mammals rather than lighter birds, it will nest lower down the mountain to avoid wasting precious energy carrying food up to the heights.

Like many larger animals, a golden eagle has few offspring. Usually two eggs are laid a few days apart, but in most years a pair will rear only one chick. Having hatched earlier, the older chick is bigger and stronger than its younger sibling and gets the lion's share of whatever food is brought back to the nest. Gradually, the younger chick gets weaker and, often after being bullied by its stronger sibling, usually dies.

So why lay two eggs? The theory is that, in good years, when the weather stays fine and food in the mountains is plentiful, both chicks can be raised to the fledging stage. At other times, the second chick is an 'insurance policy' against the first one dying unexpectedly. Scientists have called this unusual behaviour Cainism, after the Old Testament story in which Cain murdered his brother Abel. But for the eagles, it is simply a sensible precaution in an environment where nothing can ever be taken for granted.

◄ **Carrion clash.** A young golden eagle in its first winter (left) clashes with his father over access to carrion high in the mountains in central Norway. An eagle's home range needs to be huge for it to find enough food in the mountains – carrion being vitally important in winter – and though ranges may overlap, an eagle will defend its core area.

## Sleeping for survival

Scavenging is one way to get through the winter. But for some mountain creatures, there is a better alternative, one that doesn't leave them at the mercy of the unpredictable weather and shortages of food: hibernation. And the biggest animal found in the mountains, the brown bear, does exactly that.

The largest brown bears, the grizzly bears, are formidable. A big male can weigh up to 320kg (700 pounds), and the female, while roughly half its weight, is still a fair size. Bears are omnivorous, feeding on a wide range of plant and animal food, including grass and berries as well as fish

► **Rub-tree message board.** A male grizzly gives his back a good scratch and at the same time leaves a scent mark for passing bears, whether potential partners or rivals.

▼ **Grizzly chase.** On a mountain slope in Alaska, a grizzly tries to catch a ground squirrel – a valuable protein snack, especially when building up reserves for hibernation.

(especially salmon), birds' eggs and small mammals such as squirrels – basically, anything they can find. But in autumn, as daylight shortens, the reliable food supplies diminish. This is especially true at high altitudes, where winter snows may come early and with little or no warning. So the bears spend the last precious days of plenty feeding almost constantly, consuming up to 40kg (90 pounds) of food every day, building up their weight for the winter months to come.

Then in late October or November, a bear looks for a place to hibernate. It either uses a natural site, such as a cave, or a den it has dug, in which it spends a long period of dormancy, lowering its metabolism to use as little energy as possible. This enables it to survive on its fat reserves all the way through to the following spring – up to six months of inactivity.

But that's not the whole story. A bear might wake up to change its position in the den or even venture outside for a time. Some bear cubs are actually born during the hibernation period. A cub feeds on its mother's rich milk for the whole of its first year, but even so, it has only a fifty-fifty chance of reaching its first birthday.

If cubs and mothers do survive, then in the spring, as the sun rises higher in the sky each day, the plants begin to grow again and other animals emerge, so will the bears. Even then, the dangers of the winter are not over: there can be late snowfalls, making food hard to find, or even avalanches that can crush and kill everything in their path, including bears. But with luck and the mother's experience and good judgement, they will survive to feed in the lush summer meadows lower down the mountainside, where food is plentiful and easy to find.

During the short summers, the female grizzlies take advantage of the long hours of daylight to put on as much weight as they can, not just for themselves but also for their cubs. This is a race against time – the northern summers are productive but brief. Like any visitor to the mountains in summer, they have to contend with clouds of biting insects. That is one reason you often see bears rubbing themselves against trees or rocks, to gain some relief from the constant itching.

◄ **Grizzly grazing.** A grizzly feeding in a spring meadow clearing in the temperate, coastal rainforest of the Great Bear Rainforest, British Columbia, Canada. It is a time of plenty, when the bears come down to the lower levels and eat as much as possible to start putting on weight lost over the winter hibernation.

## Cold nights and warming up

You might think that if you live close to the equator, even if you are halfway up a mountain, life would be easy. But the higher the altitude, the further the average temperature drops. So in the mountains of Rwanda, which lie close to the equator but between 2000 and 4500 metres (6600–14,800 feet) above sea level, the average temperature is considerably colder than in the foothills, and at night it can drop even lower.

To combat the cold nights, mountain gorillas have developed an ingenious behaviour. First, they make a nest out of leaves and branches, usually on the ground, and then snuggle under the foliage just as we would do beneath a duvet or in a sleeping bag. But the next stage of their strategy is definitely not for the squeamish. They then defecate, the warmth of their droppings acting as a hot-water bottle.

A nest, though, is only temporary. Because they are nomadic, gorillas generally make a new one each night. But though this uses up precious time and energy, without the warmth, they – and especially their infants, which share their mothers' nests – might not survive.

Of all the different kinds of bird found in mountainous regions around the world, perhaps the most unexpected are flamingos. We usually associate these elegant waterbirds with the vast salt lakes of Africa, where half a million lesser flamingos feed on vast, open saltpans, standing in the unrelenting midday sun few other creatures can endure.

But in the Andes Mountains that run along the western side of South America, three different kinds of flamingos – James's, Chilean and Andean – have evolved to endure a very different challenge. Here, at more than 4500 metres (15,000 feet) above sea level, temperatures plummet so low at night that the birds' feet can become frozen in the ice, temporarily incarcerating them. To keep warm in these freezing surroundings, the flamingos fluff out their feathers, trapping a layer of warm air beneath.

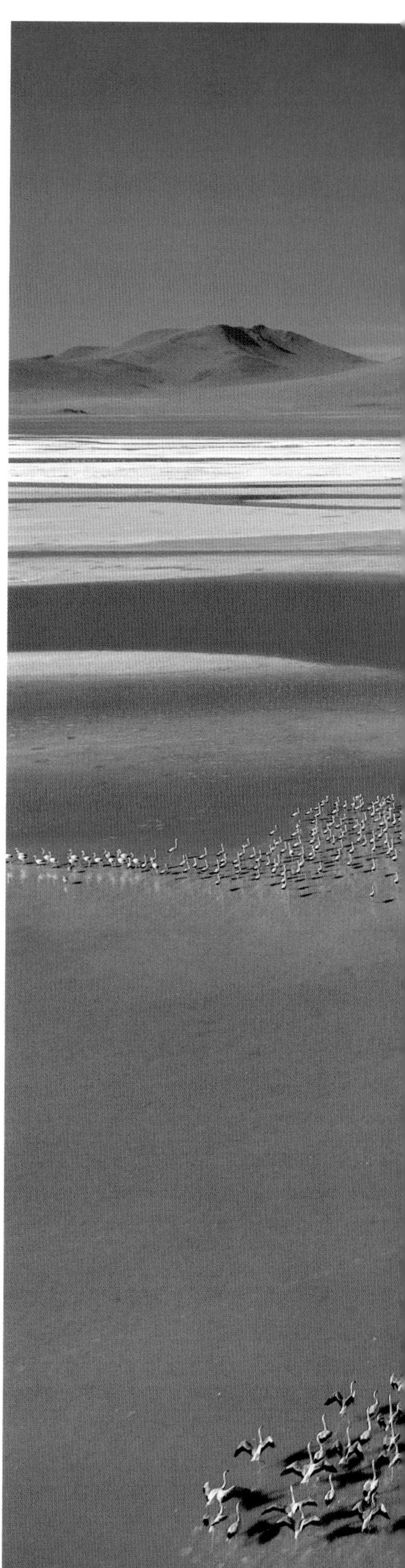

▶ **Flamingo refuge.** James's flamingos feed in the shallow water of Bolivia's Laguna Colorada, high in the Altiplano. The lake gets its name (meaning coloured) from the orange-red of the sediment and some of the algae that grow in the nutrient-rich, hot-spring-warmed water. Almost two thirds of the population of just 50,000 James's flamingos breed here. The area is so remote that the species was considered extinct for decades until rediscovered here in the late 1950s.

Others avoid the problem by gathering near the hot springs produced by volcanic activity, where the waters remain unfrozen.

The reason these birds endure such harsh conditions is because at this altitude there are few, if any, predators to trouble them; and because once the sun does begin to rise in the sky, the temperature rises rapidly. This enables the ice to thaw, so that they can escape and start to feed in the shallow, food-rich saline lakes. But how do they avoid freezing up completely? Like

other waterbirds, flamingos are able to stop the flow of blood to their legs and feet, which means that, though their lower extremities are frozen, the rest of their body stays (at least relatively) warm.

Another Andean mountain creature also needs to warm up for the day. Viscachas – of which there are several different species – are rodents closely related to chinchillas. They are about the size and appearance of rabbits, with long ears, though they have longer tails. Viscachas live

**Head-turning dance**. Andean flamingos perform a pre-courtship head-turning display on a salt lake on the Altiplano in Chile, the highest place in the world where flamingos are found. Males and females move in unison, flushed pink in their breeding plumage and eyeing up each other as mates.

◄ **Sunbather.** A southern viscacha – a large Andean rodent – warms up in the early morning sun beside its mountain den. Preyed on by eagles, foxes and wild cats, it keeps its ears tuned and eyes partly open.

► **Warm-water wildfowl-watching.** A bobcat watches intently as a goldeneye duck swims past. Volcanic springs keep the water of Madison River in the Rocky Mountains, Wyoming, USA, warm in winter, even in temperatures of -30°C (-22°F), providing hotspots for wildfowl and also opportunistic bobcats. The cat will ambush wildfowl, even leaping into the water to grab a bird.

colonially in warrens, between the tree line and snow line on the sides of the Andes, and feed on a wide variety of plants – though in this harsh landscape, food can often be hard to find. But before a viscacha can think about feeding, it has to get warm. It does this by sitting on its haunches in the early morning sun like a miniature Buddha, allowing its muscles to warm up. By mid-morning it reaches the 'golden hour', when it has enough energy to feed but when the air is still cool enough for its frenetic activity. An hour or so later, approaching midday, the sun has become too strong for it, and the viscacha retreats into the shade until later in the day, when it can resume feeding once again.

On the high tops, the weather can turn in an instant, bringing polar winds and snow one moment and baking sunshine the next; with rain, sleet, hail and strong winds as well. In winter, temperatures can plummet way below zero, and food can be even harder to find. But in some mountain ranges where there is still volcanic activity, such as in Alaska and the Rocky Mountains in North America, volcanic springs can keep the water unfrozen and produce island environments that can be crucial refuges for wildfowl and attract other animals to both water and potential prey.

▲ **Descent to water.** In the Judean Desert, male and female Nubian ibex descend to drink in a ravine – a dangerous but necessary daily event. The rocky heights where they live are a refuge against predators and disturbance from humans.

◄ **High point.** A female ibex perches confidently on a cliff face. Her cloven hooves have concave undersides that act as suction cups and help grip the rocks, and her sandy-coloured coat both helps her blend into the background and reflects the sun.

## Descending goats and climbing bears

Not every challenge faced by mountain creatures is related to heat or cold. Just as in deserts, finding reliable sources of drinking water can also be tricky. In mountain ranges in hot countries, this problem may be made even worse by the lack of rainfall and the rapid evaporation of any water that does reach the ground.

In the mountains of Arabia, the little rain that does fall usually runs straight down the arid, vertiginous slopes and collects in riverbeds, known as wadis. These are usually dry, but after rain, they temporarily fill with water, providing a much-needed resource for all the mountain creatures that live in the region.

But to reach this precious liquid, the animals have to first climb down precipitous rock and so must be expert climbers. One of the most expert is the Nubian ibex, a member of the goat family. It is specially adapted to living in dry, mountainous regions found from Algeria in the west through to Oman in the east, and into Africa as far south as Sudan. But despite this large range, there may be only 1200 Nubian ibex left in the wild.

Because ibexes live in separate male and female herds, the job of raising the young, known as kids, and guiding them down the tricky mountain slopes to find water, falls to the females. Even when the kids are only a few days old, they may have to make the journey – for if they fail to find water this time, it can be many weeks before it rains again. The mothers are sure-footed and perfectly adapted to life in this vertical habitat, but the kids are unsteady on their feet and uncertain of their destination. They are also targets for predators such as golden eagles, leopards and occasionally even lammergeiers – a species of vulture that ranges over these mountains.

We expect mountain goats such as ibexes to be expert mountaineers, but some animals – such as bears – just do not appear suited to climbing steep slopes. Yet one habitually does: the spectacled bear – the only bear species native to South America. It is now confined to a narrow strip of mountainous land in the northern part of the Andes, from Venezuela, Colombia and Ecuador in the north to Peru, Bolivia and northwestern Argentina in the south, though even here numbers are declining. It is a medium-sized bear, though males are much larger than females, weighing up to 200kg (440 pounds), compared to a maximum of just over 80kg (180 pounds) for the females. It has a broad, blunt snout and pale markings on its face, which give the species its name; these markings vary considerably between individuals, making some resemble their distant cousin the giant panda.

Found in a variety of montane habitats, most commonly in cloud forests, high-elevation elfin forests and the upper slopes of the Andes, the spectacled bear is largely herbivorous, with meat making up just 5 per cent of its diet. However, some bears will make a dangerous journey up sheer rock faces in search of a special ingredient in their diet: land snails, which live in the rocky crevices on precipitous mountainsides. Agile tree-climbers, these bears have evolved special adaptations – sharp claws, bent heels that can grip like modern climbing shoes, and longer limbs and less fat than other bears. Rather like human mountaineers, they keep at peak fitness.

As with the ibex, climbing presents few problems for an adult spectacled bear, but for a youngster, it is a risky learning process. It has to follow closely behind its mother, putting one paw forward at a time, following in her footsteps until it finally reaches the top.

► (top) **Cliff-hugging.** A mother and her youngster scale a near-vertical cliff-face searching for snails to supplement their diet. Their feet are adapted for climbing – trees as much as rocks – with claws and bent heels that help them grip.

► (bottom) **Mountain-running.** A bear (its coat covered in burrs) descends a mountain slope in the Peruvian Andes to reach a waterhole. It also feeds lower down but returns to a mountain ledge or cave to sleep.

## The highest life

Of all the world's mountain ranges, none are higher or more challenging than the Himalaya. These mighty mountains, whose name translates as 'place of snow', lie along the border between Nepal and Tibet to the north and India to the south, and also extend into Pakistan in the west and Bhutan in the east. They run in an arc from roughly northwest to southeast for 2400km (1500 miles), and at some points are 400km (250 miles) wide. The Himalaya contain nine out of the ten highest mountain peaks in the world, including Mount Everest, which reaches an altitude of almost 8850 metres (29,035 feet) above sea level. Known for centuries in Tibet as Chomolungma – meaning 'mother of the universe' – this is the highest point on the planet.

Ever since westerners made their own discovery of Everest in the mid-nineteenth century, it has attracted western climbers eager to reach its summit, most famously in 1924, when George Mallory and Andrew Irvine perished in the attempt, and then the triumphant expedition of 1953, when New Zealand mountaineer Edmund Hillary and Sherpa Tenzing Norgay finally stood on top of the world. Since then many more climbers have made it to the top, but some have tragically died trying.

Few animals can survive much above the lower slopes, because the higher you go, the less oxygen is available. Above 7200 metres (just below 24,000 feet), it becomes very hard for humans to breathe without the aid of extra oxygen. The tree line stops at about 4900 metres (16,000 feet), and though snow leopards, Bengal tigers and yaks live on the middle slopes of the Himalaya, only a handful of species get anywhere near the highest peaks. One notable exception is a bird: the yellow-billed or Alpine chough. These glossy-black crows, with their short, curved, custard-yellow bills, have been seen following mountaineers on Everest to 8200 metres (26,900 feet), just a few hundred metres below the summit. But they feed and breed at much lower altitudes. Yet one small invertebrate beats even this extraordinary feat of endurance.

▶ **Mother of the universe.** Last light on the top of Everest. At 8850 metres (29,035 feet), it is the highest peak in the world. And that makes the Everest jumping spider, found at altitudes of up to 6700 metres (22,000 feet), the world's highest-living predator.

The Himalayan (or Everest) jumping spider survives all year round on the mountain at heights of up to 6700 metres (22,000 feet), making it the highest-living resident predator on the planet. Its scientific name, *Euophrys omnisuperstes*, means 'standing above all others'.

The extraordinary spider was first discovered by George Mallory and his colleagues, at a height no human had ever reached before. At first, scientists assumed that it was a one-off observation – the spiders having drifted up on air currents from the foothills below. But later visits confirmed that the spiders do indeed live just below the summit of Everest.

So how do they survive? When the sun does come out, warming the air for a few hours at a time, the spiders emerge to search for food, and when it disappears, they conceal themselves beneath rocks. With no

▲ **Jumping tiger.** With the sun out and the thin air heating up on Everest, a jumping spider emerges from its shelter under a rock and starts to hunt. It has a limited time to catch windblown flies and springtails, which must be alive so the spider can suck out their body fluids. Its huge eyes will pick out any movement on the slopes.

other permanent living creatures up here to prey upon, the spiders' diet is totally dependent on tiny insects such as springtails, which are blown up the slopes by the constant wind.

The spiders find these windblown morsels with their excellent eyesight, which is particularly sensitive to the insects' tiny movements. And like all jumping spiders, they are able to force their body fluids into their legs, creating a higher pressure and enabling them to jump up to 30 times their own body length. But before leaping through the thin mountain air to catch its prey, a jumping spider will spin a silk safety line – the invertebrate equivalent of a climbing rope.

## The high flyers

For migratory birds, the high peaks of the Himalaya present a very different problem: how to get over this huge natural barrier. Yet two species of waterbird – the demoiselle crane and bar-headed goose – do manage to do it. Twice a year, on their outward and return journeys from their breeding grounds in central Asia to their winter quarters in India, they manage to cross the Himalaya, sometimes reaching seemingly impossible altitudes.

The bar-headed goose is a striking bird, with a bright orange bill and legs and a head marked with two black bands behind the eye, which give the species its name. They breed in vast, noisy colonies on mountain lakes in central Asia, from Russia in the north through Kazakhstan and Mongolia to China and Tibet in the south. All these countries lie north of the Himalaya, so to reach their winter homes on the low-lying fields and wetlands of Pakistan, northern India and Bangladesh, they have to cross the Himalaya.

Like most members of their family, bar-headed geese travel in large flocks, often flying in V-formation, with experienced adults taking the lead and their untested youngsters following. On these epic journeys, they have been reported over Mount Makalu – at about 8500 metres (28,000 feet), the fifth highest peak on the planet – and even supposedly over Mount Everest itself. However, the highest reliably proven sighting is at a rather lower – though still impressive – altitude of just over 6500 metres (almost 21,500 feet), through one of the gorges that allow passage across the range. In fact most bar-headed geese divert to go through these mountain passes, even when this means adding huge distances to their already long migration. But the geese that cross at higher altitudes do so through a number of physiological adaptations. They have larger wingspans and bigger overall wing areas compared to their weight than other wildfowl. This enables them to gain and maintain height more rapidly. They can also store more oxygen in the haemoglobin in their blood, allowing them to maintain the oxygen supply to their flight muscles for a prolonged period. And they can raise their heart rate, and so increase the amount of blood being pumped at any one time.

Though some migrating birds do choose to embark with a following wind to help them on their way, bar-headed geese prefer calm conditions,

**High-powered high flyers.** Bar-headed geese fly on migration through the Himalaya to winter in India. They hold the record for the greatest rate of climbing flight, powered by special physiological adaptations, though they normally fly through the passes rather than over the peaks.

which reduce their chances of being caught in an unexpected and potentially fatal storm. They mostly travel by night or in the very early morning, when cooler, denser air enables them to attain a greater height – and to stay there.

Even so, many perish along the way; either falling from exhaustion, getting caught out by unexpectedly bad weather, being shot or being hunted by aerial predators such as golden eagles, for whom this twice-yearly migration brings a bonanza of food. But enough of these extraordinary birds do make it through to maintain their population.

## Pressure from above and below

In a world where wild creatures are few and far between, there is one other problem – how to find a mate? This is particularly so for big predators at the top of the food chain, whose numbers are low and whose territory is necessarily very large. It is made worse by the vertical nature of mountain terrain, which prevents sounds and scents travelling over long distances. Chronic bad weather can also make seeing over long distances difficult.

Snow leopards have huge home ranges of up to 200 square kilometres (about 80 square miles), and are generally solitary. But when they need to

▲ **Watchout.** A male snow leopard rests after feeding, keeping watch over its kill in Ladakh's Ulley Valley.

mate – usually towards the end of the winter – they make contact using scent and sound. They urinate in strategic places or spray selected rocks to leave 'pee mail' messages for any potential mates and indicate exactly who is in the vicinity. Females also 'sing' when they are receptive, their calls echoing across the valleys. After mating, they resume their solitary lives.

Gestation lasts about three months, and the female gives birth in late spring or early summer to between one and five blind, helpless cubs, depending on the health of the mother and the availability of food.

Each cub's survival is paramount for the future of the species, as today there are only 5000–7000 snow leopards in the wild, more than half in

China. As well as being hunted for their fur – a quality pelt can be worth several thousand dollars – the animal's bones and other organs are in demand for use in traditional Chinese medicine. Where there is poverty, the temptations to poach are great.

But the biggest problem faced by this elusive creature – and indeed its high montane habitat – is climate change. As global average temperatures increase, many species will be forced to migrate upwards to cooler climes to a point where they can't go any further.

Snow leopards typically hunt on mountain slopes, in the narrow area between the end of the tree line and the start of the snow line. Now they are hunting at higher altitudes than ever before. But the higher they go, the scarcer are the creatures on which they feed.

Meanwhile, climate change is also causing desertification lower down, forcing traditional herders to move their livestock to higher altitudes. Not only do their animals compete for food with the leopards' prey, such as wild sheep and ibexes, but hungry snow leopards can be forced into killing livestock. The result is retribution killings by desperate herders. With so many threats, careful protection over the coming decades will be needed to prevent this magnificent animal disappearing altogether.

It could be argued that, of all the world's habitats, mountain ecosystems are among the least damaged. Their isolation has saved them from changes that have devastated lowland habitats such as forests and grasslands, or the damage caused by the arrival of humans and their domestic animals on once-pristine oceanic islands. But isolation is its Achilles heel. This has enabled a range of highly specialized ecosystems to evolve, each characterized by highly specialized plant and animal species. Now that this environment is changing – in many cases rapidly – through local factors such as increased human pressure from farming and hunting and most severely from climate change, the wildlife of the world's mountains faces an uncertain future.

▶ **Hunting move.** A female snow leopard in Ladakh's Ulley Valley comes down from her lookout point to a fresh kill. These cats hunt mainly wild sheep and ibexes, but if prey is scarce, they will also take domestic animals.

▶ (next page) **High-top pass.** A camera-trap captures a passing snow leopard high in Ladakh's mountains. Camera-traps are revealing more about the social lives of these elusive big cats.

# 3
# DESERTS

**DESOLATE, BARREN WASTELANDS,** where there is little or no water and even less life. Baking hot by day and freezing cold by night. Eerily quiet, the only movement being the shifting sands, blown by a constant wind. And a distant mirage that, however close you get to it, is tantalisingly out of reach. This is the image most of us have of the world's deserts. And, of course, for much of the time, deserts do fit this bleak description. Yet at other times, it could hardly be further from the truth.

Though the very definition of desert is a place with rainfall way below average, usually defined as less than 250mm (10 inches) a year, sometimes it does rain, heavily. When this happens, life appears as if by magic, and for a short while plants flourish, turning the desert green.

Nor is it always hot. Many of the world's deserts – including much of the Arctic and Antarctic, where rain rarely falls – are bitterly cold. And though deserts can be sandy, they can also be formed of rocks and stones, creating a very different and equally spectacular landscape. So while it is true that few other habitats on the planet offer so little, this does not mean that deserts are devoid of life. You just need to know where and when to look for it.

To survive in this harsh environment, plants and animals have come up with some of the most ingenious survival strategies. To conserve water, many plants have dispensed with leaves altogether and have grown spiny coverings to discourage animals from eating them. Some bloom only when rains come, which may mean they stay dormant for many years. Others

(previous page) **The search for water.** Related family groups of elephants in Etosha National Park, Namibia, trek across a dried-up riverbed. They are led by the oldest, most experienced female, who has a mental map of the region, built up over years, and will know the shortest direct route to a water source – vital knowledge for survival in this arid landscape.

◀ **Desert storm.** A summer storm in Arizona's Sonoran Desert, one of the wettest deserts in North America. Such storms bring bursts of heavy rain, which the saguaro cacti absorb through extensive roots just under the surface, their pleated stems expanding to store it.

have extensive roots that run long distances under the surface, ready for any rain, or a few may even find moisture deep under ground.

The majority of desert creatures lead a nocturnal existence, often hiding away during the daylight hours in burrows beneath the sand or rocks and only emerging at dusk to feed. Some species also delay starting their reproductive cycle until the rains come, but then, while the conditions are beneficial to raising a family, they breed as rapidly and frequently as possible.

All these adaptations allow desert creatures to thrive in a dry, desiccated world where life's challenges are centred on getting the one thing that no living creature can do without: water.

## From drought to deluge

What unifies all the world's deserts is a lack of water, but that doesn't mean that in some deserts, at certain times of the year, rain won't fall. Each year in the deserts of the southwestern United States, billowing banks of rain clouds form on the horizon, but the rain they bring is not always welcome. For this is the monsoon, not perhaps as spectacular as its Asian counterpart, but at times just as dramatic.

For a couple of months, from midsummer until early autumn, each afternoon, as the heat reaches its peak, rain clouds form massive thunderheads and, accompanied by dazzling flashes of lightning and deafening crashes of thunder, unleash countless tonnes of water on the desert, changing the landscape from drought to flood in minutes.

For the creatures that live in these deserts, too much water can be even more of a problem than too little. With nothing to hold onto the water, destructive flash floods tear through these rocky canyon-lands, ultimately shaping the landscape itself. Far from being the source of life, so much water rushing through the land in such a brief period is an apocalyptic force of destruction.

▶ **Desert deluge.** In the Namib Desert in southwest Africa, a short but intense storm dumps a heavy fall of rain on one small area, causing flash floods. A strong downdraught, caused by huge drops of rain pulling the surrounding air down with them, has produced a groundcover of rain spreading over the normally parched land.

▲ **Fast feeders.** Locusts in their gregarious, swarming phase rapidly consume an area of grass in Madagascar before flying on to find a new place to feed. A population explosion after desert rain and a glut of food, followed by a new shortage of food as drought conditions resumed, caused metabolic and behavioural changes that have transformed the normally solitary grasshoppers into a swarming horde.

◀ **Desert swarm.** A tiny part of just one of the locust swarms that developed in Madagascar in 2015. This particular swarm contained several billion individuals, consumed an estimated 40,000 tonnes of vegetation each day and covered more than 520 square kilometres (200 square miles).

## From famine to feast

There are some desert plants and animals for which a sudden downpour and the resulting floods are not a curse. The water may only stay on the surface for a few hours, but if it disappears into the ground, it can still be accessed – by plants. So when the rains fall over the deserts of Africa, seeds that have lain dormant in the ground for months burst into life. This is the cue for another spectacular explosion, of animal life.

Locusts are a group of short-horned grasshoppers, which are normally solitary. When food becomes temporarily abundant, there is a population explosion, but as the food runs out, they are forced together, and close proximity triggers a chemical change in their brains and a physical change. They now form vast, nomadic swarms that fly in search of new desert grasslands on which to settle and feed. In the developing world, they can cause massive damage to fragile agricultural systems. Such locust plagues were mentioned in the Old Testament of the Bible and also appear on Egyptian tomb carvings.

The locusts have evolved to be quick off the mark to take advantage of the glut of food, as it may be years before such riches appear again. First a few pioneers appear, quickly followed by more and yet more locusts, until hordes darken the skies, settling to exploit any new green pastures. Each swarm covers huge areas and consists of as many as a billion locusts. So it is hardly surprising that they devour everything in their path as they fly in waves across the landscape at speeds of up to 20kph (12mph). Almost as soon as the desert has turned from dusty yellow-brown to green, these vast swarms of insects lay it bare once again.

► **Saguaro lookout.** In Arizona's Sonoran Desert, a Harris's hawk sits on a lookout post, a saguaro cactus, watching for prey. Hunting with it as a pack are two other hawks, all looking to see if they can flush out a jackrabbit or pack rat. Harris's hawks are the only birds of prey known to hunt cooperatively, which makes sense in a desert landscape where there are few prey animals and lots of spiny cover for them to hide in.

## Game of thorns

One particular family of plants, the cacti, has evolved to hoard the precious liquid and so overcome the feast-or-famine nature of desert rainfall. There are about 1500 species native to the deserts of North and South America and one species native to Africa, but as popular plants, they have been introduced across many other countries. By far the largest and most impressive of all is the saguaro, the classic cactus of the deserts of New Mexico and Arizona. Featured prominently in countless Western films, it can live for 150 years and grow up to 20 metres (65 feet) tall. This giant also has a clever trick. When there is a downpour, it is able to soak up as much as a tonne of water in the following 24 hours. Having done this, the saguaro then has to protect its booty against any creature that might wish to exploit this valuable resource. It does so with a barrage of long, sharp spines over almost its entire surface.

Many of the desert's smaller inhabitants have turned the saguaro's spiny defences to their advantage, using the cacti as fortresses to keep them safe from predators, either living within them or fleeing to the safety of the spiny 'forest'. But one very clever predator has learned to get around these defences. Harris's hawk is a large, long-winged, desert-dwelling raptor that hunts cooperatively, rather like lions. First, a scout perched on a saguaro spots the prey, perhaps a jackrabbit. Then one or two birds fly low towards the animal, causing it to run for cover. Having reached what appears to be the safe haven of cactus spines, the jackrabbit might think it has managed to escape, but the hunt is far from over. Now the ground crew move in, stalking through the forest of cacti, trying to flush

▲ **Butcher-bird larder.** In a desert area of New Mexico, USA, a male loggerhead shrike removes food from its larder – yucca spines on which it has impaled lizards – and takes the corpse to feed to its chicks. The spines will hold firm larger prey that it needs to butcher, though it usually kills with a bite to the back of the neck.

◄ (top) **Hunting duo.** Two Harris's hawks swoop down after a jackrabbit, navigating though the maze of cactus spines at surprising speed as their potential prey races for cover.

◄ (bottom) **Ground search.** While the other hawks hunt from above, one lands and walks through the bush, trying to flush out the prey. No matter which hawk makes the kill, the food will be shared between all of them.

the animal from its hiding place. If a jackrabbit stays put, it will probably escape. But if it panics and makes a break for it, the watchers will strike, plummeting down from their vantage points on the top of the saguaros, twisting and turning through the spiny maze until one invariably grabs the animal with its lethal talons.

Harris's hawks are the only one of the world's 300 or more species of day-flying raptors to hunt cooperatively, all in response to the spiny saguaro cactus. It's this ability to cooperate that makes Harris's hawks popular among falconers, being easy to train and quick to learn.

Another, much smaller predator, the loggerhead shrike, also takes advantage of the saguaro's spines. It uses them to hold and store its victims. It is a songbird but with a fearsomely hooked bill and a preference for small birds, large insects, amphibians, reptiles and rodents. Like other shrikes, it frequently impales its larger victims onto thorns and spikes, sometimes to dismember them more easily but also to store them. This habit has earned it the name butcher bird – an apt description.

## The big drinkers

All life in the desert needs water. For some smaller creatures, the food they eat contains just enough liquid to sustain them. But for larger ones, such as the vast herds of grazing animals that roam across many of the world's deserts, this is not an option: they have to drink. That is why, in the Kalahari and Namib deserts of southwest Africa, huge herds of wild beasts migrate to and fro across the parched land, with just one thing in mind – getting to the next waterhole.

African elephants are the world's largest land mammal – and also by far the thirstiest. Even though desert elephants are on average smaller and lighter than their grassland cousins, a mother elephant still has to consume more than 200 litres (more than 40 gallons) of water every day if she is to provide enough milk for a young calf. She also needs water

► **Desert masters.** An elephant and her calves travel along a dry riverbed in search of vegetation. Their population is adapted to live in the desert in Namibia's Kunene region. They can be smaller than savannah-living elephants, possibly because of their diet, and their feet appear wider, probably as a result of walking distances over rocks and sand.

▼ **Mother care.** A mother keeps her baby shaded by her shadow as they travel in the early morning. These desert-adapted elephants can survive several days without drinking, but not if they are suckling calves.

for herself and cannot go more than three days without drinking, which makes finding a source of fresh water essential.

Elephants may travel several hundred kilometres in the dry season, moving from place to place, led by a matriarch – the group's oldest, most experienced female. She has to build mental maps from previous journeys, enabling her to remember where sources of water have been found before and to navigate over the often featureless terrain. Sometimes the water the matriarch needs to find is not on the surface but under ground. She will rely on her acute sense of smell to find these hidden sources and then dig with her powerful feet to reach them. These skills are essential if she and the group are to survive. But just because an animal is able to find water doesn't mean that its problems are over. Female and baby elephants may be forced back from a potential source of drinking water by aggressive herds of young males and may as a result have to travel far farther to find a new place to drink undisturbed.

In the Great Basin Desert of the southwestern USA, between the Sierra Nevada to the west and the Rocky Mountains to the east – the largest continuous area of desert in the country – dwindling supplies of water cause another animal, not normally associated with desert life, to do battle. Mustangs are the classic horses of the Wild West, and herds eke out a living in these arid lands. They are descended from domestic horses that were deliberately or accidentally released. (Mustang derives from a Spanish word meaning 'an animal that strays'.) Over time these herds have lost any signs of domesticity. A male mustang is usually accompanied by a harem of females, each of which also has to drink its fill at a crowded waterhole. It is towards the end of the dry season, as the waterhole gets smaller and smaller and the dominant stallions come into contact, that the trouble starts.

The fights are short but violent, with plenty of rearing up, kicking and biting. They can also be lethal: a stallion kicking at the wrong angle might break his opponent's leg or jaw, leaving him to die a lingering death in the unforgiving desert heat.

◄ **Waterhole friction.** Two stallions, each with a harem of mares, fight beside a waterhole. In Nevada, USA, water is what limits the numbers of feral horses living in the desert areas, and in the breeding season, when groups gather to drink, access to females fuels disputes, which can escalate into dangerous fights.

## Waterhole tactics

▲ **Dangerous drinking.** A pale chanting goshawk charges a flock of Namaqua sandgrouse drinking at a waterhole in the Namib Desert. These desert-living goshawks don't need to ambush their prey. With sandgrouse desperate to drink, the hawks merely wait by the waterhole and then choose their moment.

Finding water in the desert and then actually getting enough of it is even more difficult for those creatures that are unable to travel, such as young birds in their nests. Sandgrouse are, as their name suggests, one of the quintessential desert bird families, with 16 species, found in Africa, Asia and southern Europe – all long-tailed, pigeon-like birds with camouflage-coloured and patterned plumage. Namaqua sandgrouse live in the deserts and semi-deserts of southern Africa. Like other sandgrouse, they are gregarious, travelling in large flocks in order to find new sources of food and water. They choose to lay their eggs in shallow scrapes on the ground in the desert, far away from most predators – which also means far from water. But they have an ingenious solution to the problem of access to water for the chicks when they hatch.

Each dawn the males fly in large flocks to the nearest waterhole. While they are drinking they soak up the water in their specially thickened

▲ **Father's feather sponge.** At its nest scrape, in the heart of the desert, a Namaqua sandgrouse chick sucks water from its father's special belly feathers. He soaked his sponge-like plumage while he drank at a waterhole many miles away – each feather taking up many times its weight in water.

breast feathers. These act as a sponge, holding up to 20 millilitres of liquid – enough to satisfy their thirsty chicks.

But in doing so, they put themselves at great risk, for waterholes naturally attract predators, for whom a flock of plump birds landing by the edge of the water provides the ideal opportunity for breakfast. Pale chanting goshawks, slender, long-legged birds of prey with slate grey plumage and bright red bills, hang around on the edge of the waterhole and wait. They don't usually bother to hide as they know that, before the sun rises high in the sky, the sandgrouse will have to come down to drink.

As soon as the flock has arrived, the goshawks take their chance, flying low and fast across the ground. But sandgrouse know that they are vulnerable when drinking and keep watch as a flock for movement. Usually one sounds the alarm, and they manage to take flight before the goshawk reaches them. If a sandgrouse is caught, it's usually a lone straggler on the edge of the flock – food for the goshawk's own family of youngsters, safe in a nest off the ground somewhere nearby.

## Barrier deserts and life-saving oases

Sandgrouse have adapted to live in the desert, but for most birds, this is an alien habitat. Yet twice a year, tens of millions of songbirds have to cross one of the biggest deserts of all – the mighty Sahara – on their way from their breeding grounds in Europe and Asia to their winter quarters in Africa, and back again.

For many, especially those on their very first migratory journey in the autumn after their birth, this is one barrier too many. Having left the safety of the nest and having managed to evade predators such as sparrowhawks and cats for the whole of the summer, they now have to head south on an epic journey, sometimes for many thousands of kilometres. Before they leave, smaller species such as warblers, chats and flycatchers feed constantly, to put on the weight they need for the journey. This may involve almost doubling in weight, by forming a layer of fat just beneath their skin.

Once they have done this, they wait for suitable conditions to depart: ideally clear skies and light winds. Like most songbird migrants, they set off at night, when the air is cooler and allows faster flight, and when they can avoid diurnal predators such as hawks and falcons.

Their first hurdle is to cross the continent of Europe, followed by the Mediterranean Sea. Then comes the biggest barrier of all, the Sahara. This presents a mighty obstacle, for the birds are unable to fly across in a single night. That's why places such as the oasis at Merzouga in Morocco are vital. Date palms cluster around small pools of water, offering respite, sanctuary and above all places to drink and feed, all of which are crucial if these tiny birds are to survive their arduous desert crossing.

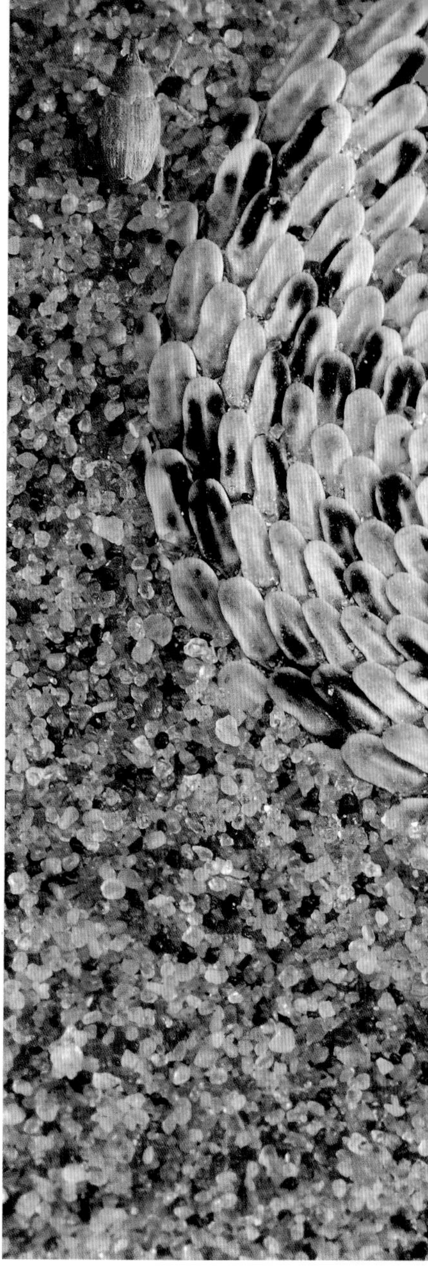

▲ **Sand sleeper.** A Peringuey's adder waits for prey, lurking just below the sand in the shaded slipface of a dune. Having eyes on the top of its head allows it to see while partially submerged. In the heat of the day, it buries down, emerging if it is overcast or at dusk to sidewind over the dunes in its search for lizards and rodents. It bites its prey, injecting venom, then holds it until it is incapacitated and can be swallowed.

## The murderous heat

A lack of water is not the only thing that can kill a desert animal. Temperatures can rise above a sweltering 50°C (122°F) – too hot for any living thing to survive without taking cover. As the sun reaches its zenith, any creature out and about must have a strategy to deal with the heat, whether limiting the time it is out by moving quickly or keeping its body well off the ground and its feet in contact with the sand for as short a time as possible.

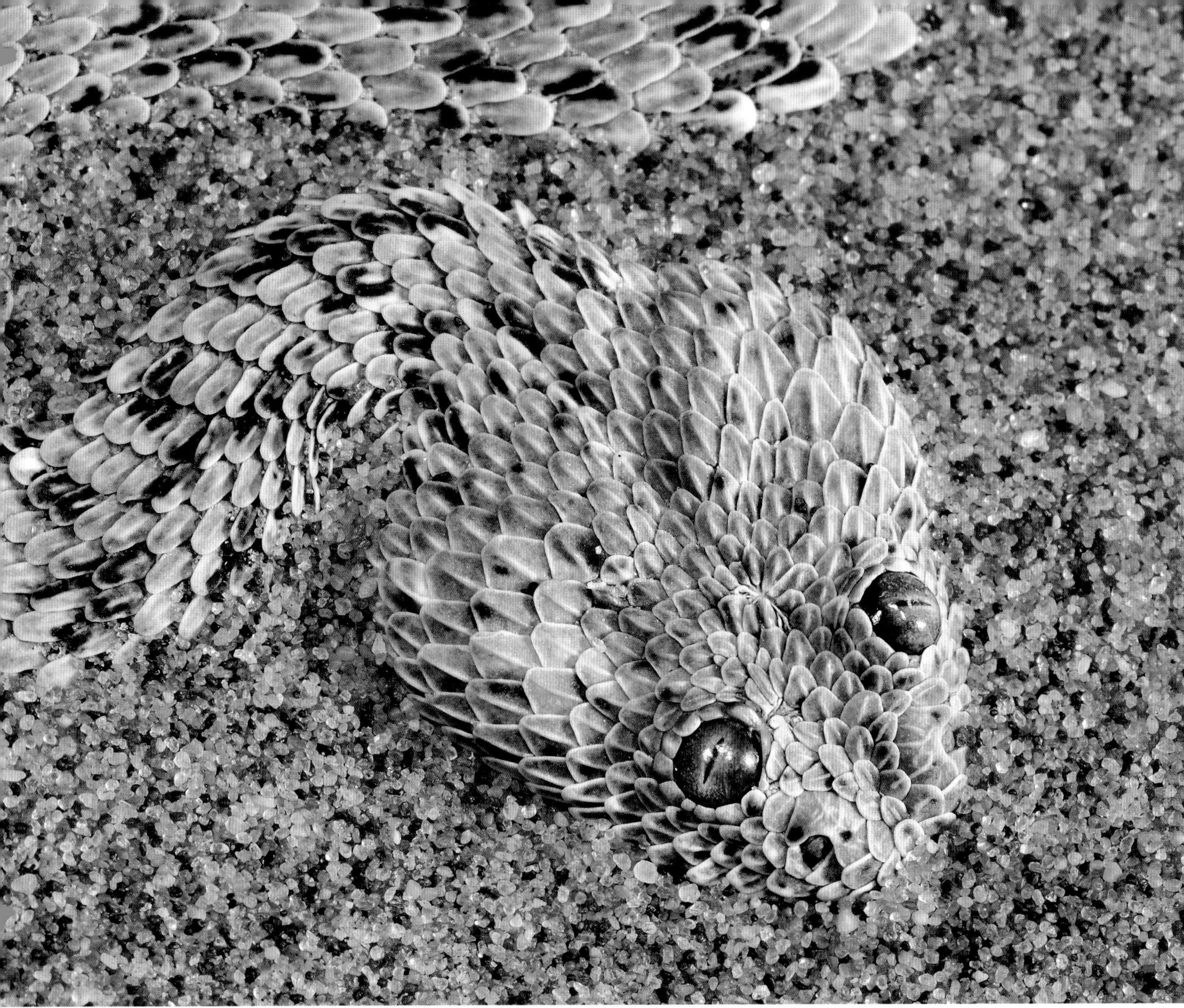

▶ (next page) **Virtual storm.** Looking like an approaching rainstorm, a haboob sweeps in at the edge of the Sahara in Sudan. It has been caused by precipitation (mainly ice crystals) from a cloud that have vaporized before reaching the ground, and the rapid descent of extremely cold air has kicked up the sand and sediment into a storm.

Snakes that are desert specialists, such as the sidewinder of southwestern North America and Peringuey's adder of Namibia and Angola, may hide under a layer of sand and simply wait for victims – usually a lizard or small rodent. As one passes by, the snake erupts from its hiding place, grabs the animal and retreats to a cooler, shaded spot to digest the meal at leisure.

In the period between the relatively cool dawn and the baking heat of noon, the ground heats up, generating movements of air that create a relentless wind which scours the desert sands. In extreme cases this can lead to massive sandstorms that sweep across the land. Tiny grains of sand bombard the rocks, scouring them into their distinctive shapes. Yet once the sun sinks in the western sky, the storm drops and the rush and roar of the wind gives way to an eerie quiet.

## The flying scorpion-killer

One simple way to avoid the sun and wind is to come out only at night. And many desert creatures choose to do just that – to lead a nocturnal, or in some cases crepuscular, existence, emerging at dusk and retreating to their lairs as soon as the sun comes up. In the Negev Desert in Israel, as the sun dips below the horizon, and the stars begin to wheel overhead in the night sky, a whole suite of amazing creatures emerges to go about their lives. In this dark world, hearing and touch become important in finding your way around, looking for food and avoiding being eaten.

Yellow scorpions 'hear' with their feet, using them to detect vibrations and pinpoint prey with devastating accuracy. Any smaller insect or other invertebrate, including spiders, may be on the menu, dismembered with the scorpion's powerful claws. But these predators, too, have their nemesis.

Out of nowhere, just as the yellow scorpion is polishing off its meal, a desert long-eared bat swoops down to attack, taking advantage of the scorpion's inability to hear predators if they attack from above. The scorpion does have a defence: its sting is the second most venomous in the world. Yet the bat doesn't even bother to pick up its victim by the tail to avoid being stung, which suggests that this particular bat may have become immune to scorpion venom. The huge advantage for the bat is that a scorpion is far larger than its usual prey of small flying insects, providing a far more nutritious meal for the same expenditure of energy.

The bat has a special hunting method, switching its echolocation technique from the usual 'screaming' method, which works when catching fast-moving insects in flight, to a quieter, 'whispering' method when tracking a scorpion that is on the ground. This gives the bat a huge advantage in the tough environment of the desert, where if something moves, and you can catch and eat it, you probably should.

But the scorpion does have a chance to escape. If it stops moving and stays completely still, the bat may miss its target and be unable to locate the scorpion on the ground. Then it has to take off and start the hunt again.

► **A sound meal.** Having located a scorpion by sound – first from above using 'whispering' echolocation, then on the ground by straight hearing – a desert long-eared bat has killed and eaten most of it, head first, in minutes. It now prepares for take-off with a burst of echolocation.

## Night tactics

All bats navigate and hunt in the dark using echolocation – bouncing sound off objects and listening to the echoes. The desert long-eared bat has huge ears that act like parabolic reflectors, magnifying and intensifying the sounds so it can pinpoint even fast-moving flying insects. Locating prey on the desert floor can be more difficult and only worth the trouble if it is a large meal such as a scorpion. But a yellow scorpion can sense the bat echolocating and has a defence – staying still. Only if it moves and makes a sound will it be caught. As each animal tries to outwit the other, the hunt can go on for some time.

**1 Attack approach.** A desert long-eared bat 'hears' the scorpion and drops down, stopping echolocating, presumably so that the scorpion does not sense it coming. If the scorpion stays still and makes no sound, the bat will not be able to find it and may even walk over it.

**2 Targeting.** Walking on its wrist joint, its huge ears focused on the noise of the moving scorpion, the bat is about to strike. The scorpion has betrayed its exact location by rushing in to attack, stinger up, pincers out.

**3 Strike and miss.** The bat misses on the first bite, but with the scorpion continuing to attack and reveal its position, the battle is already lost. Striking the bat's face with its stinger will have no effect.

**4 Shake and bite.** Having grabbed the scorpion, the bat shakes it vigorously, probably to prevent the pincers getting a hold or its sting attaching, though it appears immune to scorpion venom and will eat the poison gland.

## The dune shark and the water magnet

▲ **Sand listening.** Head down, a Grant's golden mole listens for the movements of its preferred prey, termites. It hunts at night using its acute hearing, and with no need of vision, is completely blind.

► **Sand ploughing.** A Grant's golden mole hunts in the soft, shifting sand of a dune crest, alternating between swimming just beneath the sand and ploughing over the surface, putting its head into the sand every few steps to listen and take a bearing. At sunrise, it will bury down into the dune and wait out the day in a torpid state.

One predator in the Namib Desert is so tenacious that it has been dubbed 'the dune shark'. Yet Grant's golden mole is not only small – a maximum of 9cm (3.5 inches) long – it is also blind, relying on its acute sense of hearing to hunt down prey, chiefly termites, and is adapted to survive on just the water it gets from its food. It's not related to moles, the similarities being the result of convergent evolution – adaptation over time to similar environments.

Grant's golden mole has powerful front claws for digging into the sand, and it also uses its wedge-shaped snout. But unlike its relatives, Grant's golden mole doesn't dig a proper burrow system, probably because it lives in shifting sand in the sand dunes. But it burrows down to shelter from the daytime heat, emerging at dusk to feed. Then it swims

just under the surface or ploughs over the top listening for the vibrations of juicy termites or other small prey.

Another group of creatures in the coastal Namib Desert have come up with an ingenious solution to the challenge of finding moisture in a desert. Fogstand beetles obtain their water out of the air. At dawn, sea fog sweeps in across the coastal desert. It forms when the warm, moisture-laden air coming in over the sea from the southwest is cooled by the coastal cold Benguela current, leading to a layer of cold air overlain by warm air

**Fog-baskers.** As fog rolls in over the Namib Desert, head-stander beetles take up a characteristic fog-basking pose on the ridge of a sand dune. They will drink the moisture that gathers on their wing cases and runs down to their mouthparts – a scarce but predictable source of water.

(next page) **Dune sea.** Sea fog rolls in over the coastal dunes of the Namib Desert bringing life-giving moisture.

blowing over the desert. The result is fog. This will burn off as soon as the sun begins to rise. So before sunrise, a beetle will climb to the top of a sand dune, angle its body towards the oncoming breeze and simply wait. Tiny droplets of moisture from the fog condense on its body and run down special grooves towards its mouth, so that it can drink the precious liquid. The beetle's hard carapace has both hydrophobic and hydrophilic sections, the first repelling water, the second attracting it, which means that it can maximize the amount of water it collects and then control where it goes.

## Desert hunters

When we think of lion habitat, the first image that comes to mind is the savannah grasslands of East Africa. But in Namibia, there are lions living in the desert – a special population, adapted to survive in this dry and very demanding environment.

The Namib Desert stretches along the Skeleton Coast for more than 2000km (1200 miles) and inland for hundreds of kilometres. In the past, lions were relatively common in the coastal regions, the mountains and the desert between the ephemeral (transient) rivers. There are also records of

▲ **Desert pride.** One of the small lion prides that survives in the Namib Desert. They are crossing the dunes on their way to an oasis in search of prey. Oryx, springbok and ostrich form the main part of their diet.

lions foraging along the coast, hunting the fur seals and seabirds that breed there and scavenging – even on the occasional beached whale. But today, the remaining prides mainly hold territories farther inland between the river systems, where there is more chance of larger prey, such as oryx, zebra and springbok, and even giraffe, as all desert mammals must regularly come to drink. And like all lions, they are opportunistic and will hunt both in daytime and at night, though they usually rest in the heat of the day.

Though the desert is largely unpopulated, people live on the outskirts, which is where there are increasing conflicts between lions and farmers, who trap, poison and shoot the lions. And though ecotourism is bringing more and more visitors to see these extraordinary lions, much of the revenue generated does not filter down to the local communities, causing resentment. Trophy hunting is also legal.

To discover more about the lions' desert lifestyle and find ways to prevent conflict with humans, scientists have been putting radio collars on some lions. One celebrated male was followed for more than two years, during which time he wandered around an area covering twice the size of Wales. He also walked almost 13,000km (8078 miles), equivalent to wandering the whole length of Africa from north to south – and back. No other lion has ever been found to have travelled so far in such a short time. But in August 2014 he was shot and killed, a fate increasingly common among lions throughout the African continent, and the reason that the species is now globally threatened.

► **Epic search.** The pride walks through a massive dune belt, in search of their missing youngsters. Desert lions have exceptionally large home ranges, in which it is easy to lose a pride member.

▼ **Brotherhood.** Two of five young males in the pride chase each other in play – valuable for bonding and honing hunting skills. Unlike females, males are usually only playful until they reach adulthood.

# Race for life

Namibia's desert-living lions occasionally risk hunting very large prey, including adult giraffes, which can be 5.5 metres (18 feet) tall and weigh up to 1360kg (3000 pounds). In fact, giraffe-hunting has become a culture in at least one pride, with the females teaching their male offspring how to adopt specific roles in a formation to corner their quarry and bring it to the ground, where it can easily be killed. With large prey in open desert conditions, the odds are always against the lions – and there is the danger of being injured by a kick – but depending on the terrain, their success rate improves as the group size increases.

**1 The chase.** Two lions chase a giraffe to the opposite side of a dry riverbed, where a third lion is lying in wait. Cooperative hunting is key for desert-living lions.

**2 The strike.** The oldest lioness of the pride forms the ambush, and courageously makes a leap for the giraffe's neck.

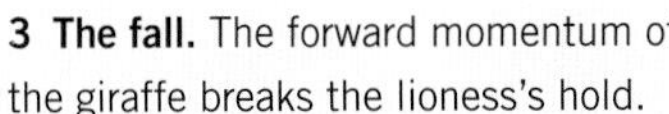

**3 The fall.** The forward momentum of the giraffe breaks the lioness's hold.

**4 The kick.** As the lioness falls, the giraffe delivers several kicks – blows capable of breaking bones. But the lioness escapes unharmed, as does the giraffe.

▲ **First shoots.** After the first winter sea fog sweeps up the dry valley in Peru's Lomas de Lachay reserve, a transformation begins. Within a week, the seemingly dead mito trees begin to revive and seeds lying dormant on the desert floor germinate.

## Paradise found, temporarily

The tenacity of desert plants and animals, and the many ingenious ways that they have learned to exploit the benefits of living in this harsh environment has to be admired. But sometimes hardly any effort is needed if you get your timing right. When certain conditions occur, usually only once every few years, deserts can explode into life.

In Peru, in the coastal foothills of the Atacama Desert, the driest place in the world, lies Lomas de Lachay. This arid land is transformed by fog – not regularly, as in the Namib, but during the mist season, from June to November. What is desert for the rest of the year becomes a verdant paradise. This newly lush landscape attracts hordes of birds, mammals

▲ **Desert meadow.** Lush growth now covers the Lomas de Lachay desert valley, watered by the fog rolling in from the sea. Within a month it will be a meadow paradise for birds, bats and insects, and the trees will flower and produce their papaya-like fruit.

and insects eager to take advantage of the annual bonus of food. The moisture from the fog also supports a massive bloom of vegetation, including trees, succulents and mosses, whose foliage drips constantly with water condensed out of the air. Gaudy hummingbirds buzz from flower to flower drinking nectar, while flocks of noisy mountain parakeets screech overhead like squadrons of fighter jets.

Lomas de Lachay is not the only place where seasonal changes bring life to the desert: the Australian Outback, the Kalahari Desert and many other arid places across the globe occasionally enjoy the benefits of rain or fog bringing precious moisture. But even when the rains don't come, for those creatures that are adapted to find water in the desert or to live without it, these special places are not somewhere to endure, but where life can thrive.

# 4
# GRASSLANDS

**ONE QUARTER OF THE LAND** on our planet is covered with grass. From the vast plains of the African savannah to a flower-rich hay meadow in the British countryside, across the pampas of South America and the steppes of Siberia, grasslands are some of the most productive habitats on Earth. They are also the most widespread, found on the edge of deserts, on the tops of mountains and in the middle of rainforests – even in coastal areas covered twice daily by the sea.

The reason grasslands are so widespread is that grasses are among the toughest and most adaptable plants. Their rapid expansion over the planet began about 8 million years ago when the climate changed and the forests started to dry out. Today, they can be as short as the surface of a billiard table or taller than an elephant. They can cope with fire and flood, grazing and trampling, rain, snow and the merciless tropical sun. They have long served human needs, too, providing food (for us and for our livestock), drink (grain is a key ingredient in beer and whisky), building materials, paper and fuel.

This is a world where change is constant, rapid and relentless. To survive, every grassland creature, from the mighty elephant to the tiny harvest mouse, must adopt a lifestyle that can cope with change. Grasslands also become the stage for some of the greatest dramas in the natural world. The battle of wits between predators and prey is the most visible spectacle, but the dramas of changing weather and seasons and the hidden battles to survive them are also played out on these grassy arenas.

To understand such a dynamic world, it helps to experience the grasslands through the eyes of the animals that live here. Take a lion cub. It

(previous page) **Grassland predator.** A mouse-eye view of an aerial predator that feeds primarily on grassland rodents – a barn owl, hunting above a meadow in Somerset, UK.

◀ **Grassland consumers.** Cape buffaloes move across Botswana's Okavango Delta in a landscape transformed by the seasonal rains into rich green grazing. They are accompanied (top right) by cattle egrets, feasting on the insects that the buffaloes stir up from the grass.

is born into a veritable paradise: a verdant swathe of green with an all-you-can-eat buffet available on tap. But within its first year, its grassland home will go through dramatic upheavals. Rains will bring floodwaters, the sun will bake the land bone-dry, and lightning will set dry grass ablaze. How it faces these challenges will dictate whether it lives or dies. The same is true of every other wild creature that lives on these vast open spaces.

► (top) **Grasstop view.** Even an Indian elephant struggles to see over the tall elephant grass that covers much of the vast floodplain of India's Kaziranga National Park.

► (bottom) **Danger trail.** In the early morning, a matriarch leads the family group through the elephant grass labyrinth on the way to water, eating as they go. A tiger could be lurking in the grass, and so the mothers keep their young calves close to them.

## The elephant, the tiger and the grass

Most of the time, you would expect one of the largest land mammals ever to roam the Earth to be able to cope well with grass. But when you're a baby Indian elephant, and the grass is more than 6 metres (20 feet) high, there can be a problem, especially if one of the world's most dangerous predators – the tiger – is lurking nearby.

Elephant grass is the world's tallest grass. The place to see this magnificent plant is in India's grassland kingdom, Kaziranga National Park, in Assam in the far northeast of the country. The park is a precious island mosaic of wet savannah, damp marshland and tropical broadleaf forests, crisscrossed by four major rivers, including the Brahmaputra, and designated a World Heritage Site because of its extraordinary variety of wildlife and globally important populations of key species.

Kaziranga boasts more than 70 per cent of the world's Indian (greater one-horned) rhinoceroses – more than 2000 – and has a higher density of Bengal tigers than any other protected area in India, possibly anywhere in the world. There are thriving populations of wild water buffaloes and swamp deer (barasingha), and about 1300 Indian elephants – again, the world's largest concentration.

The predominant vegetation of its damp floodplains is elephant grass, and this grassland habitat is as threatened as any tropical forest. Hidden within it lives the world's smallest pig, the pygmy hog, which builds its nest out of elephant grass. With possibly fewer than 150 individuals left, all in Assam, this tiny pig is now on the brink of extinction, almost entirely because of the loss of its grassland home.

The key to the fertility of Kaziranga is the annual monsoon flooding. While this causes problems for animals in the short term, it maintains the park's dynamic ecosystem. Inundation stops scrub and trees from

◀ **Grass stripes.** A tiger moves stealthily through dry elephant grass in India's Kaziranga National Park. The striped coat provides this ambush hunter with the perfect camouflage.

▶ **Elephant grass.** Using the extraordinary navigational skills handed down the generations, a matriarch elephant brings her family to the water's edge.

colonizing the land, and the river water causes the grasses to regenerate, providing an annual crop of grass for the grazing animals. They in turn attract Asia's top predator, the tiger.

At the onset of the monsoon rains, usually in June, the park's elephants head south towards the hills of Karbi Anglong. In autumn, after the rains have come to an end and the waters begin to recede, they head back to Kaziranga, where the new grass provides bountiful food. But for the baby elephants encountering this habitat for the very first time, there is a problem. They are prone to dawdling. Here it can mean becoming separated, and by the time its mother has realized her baby is missing, it may be in mortal danger.

Elephant grass provides perfect cover for an animal that is a stealth hunter and has the superb camouflage of a stripy coat. Though prey densities may not be as great as in the open grassland or damp woodland, here a hunting tiger can get close without its victim even being aware of its presence. Though the park's tigers usually prey on smaller animals such as sambar and swamp deer, elephant calves are definitely on the menu. Fortunately, a baby elephant has one very handy defence mechanism: its voice. Usually the loud distress call of a disorientated and panicking youngster enables the mother to locate it and return it to the safety of the herd.

## The swamp cats

When grassland seasonal change is rapid, it can be potentially catastrophic for wildlife. Nowhere is this more dramatic than in Africa's Okavango Delta, in Botswana. Once a year, the Okavango River surges across this flat, grassy plain, bringing almost 10 million tonnes of water with it. Within weeks, much of the area is under water, and for a few months, before the water evaporates or is taken up by plants, every wild creature here needs to re-adapt to a very different way of life.

As the waters spread across this floodplain, dormant grasses burst into life from beneath the soil. That's why up to 200,000 large grazing animals throng to the delta, ranging from elephants to buffaloes, wildebeest to rhinoceroses, and a host of antelopes, including a true wetland specialist, the lechwe. With these grazing animals come predators, including leopards, cheetahs, hyenas, wild dogs and, of course, lions.

For the lions, hungry and thin after the long, dry months when prey has been scarce, this is a glut of food they must take advantage of. So they become 'swamp cats', tracking their prey through potentially treacherous waters. It's tricky: they struggle to remain unseen and unheard, and when the chase is on, they cannot run as quickly as on land. Hidden depths bring the danger of drowning, and there are crocodiles lying in wait beneath the shimmering surface. But the sheer abundance of prey means the odds are still stacked in the lions' favour. Thus the great game of life on the grasslands continues, driven as always by seasonal change.

► **Lion crossing.** A lion crosses the Savuti Marsh in Botswana's Okavango Delta. A male can be a valuable asset for a pride hunting large animals such as buffaloes in this flooded landscape. Only a male has the weight and bulk to pull down a bull.

## Strategic teamwork

In the wet season, when the grasslands of Botswana's Okavango Delta are flooded, the lions have no choice but to enter the water if they want to exploit the herds of buffalo and other grazing animals. This may simply involve swimming to get to islands where their prey might be feeding. But some prides in northern Botswana have learned to hunt in the water itself, specializing in large and potentially dangerous species, which requires impressive teamwork. So one group now concentrates on hunting hippos, while others focus mainly on buffaloes and elephants, all of which spend much of the wet season feeding in shallow water.

**1 Circling and spinning.** The female-only pride circles a buffalo, spinning it while the lead lioness looks for an opportunity to attack its rear. Bringing down such a big animal takes the skill and experience of a team, especially if the pride is without the power and weight of a male.

**2 Rear attack.** In this foot-deep water, the lionesses find it hard to gain traction in the mud and run a real risk of slipping and being gored by the buffalo.

**3 Hanging on.** The lead lioness has dug her claws into the buffalo and is biting its spine while her sisters worry it from the front, but the fight is far from over. One false move and she could fall and be trampled. As it is, her back feet are slipping and she is in danger of losing her grip. The cubs watch and learn.

**4 Defeat.** Despite her efforts, the leading lioness has not got the strength and weight to tip the buffalo over and lets go. After almost an hour, the fight ends in defeat for the pride.

▲ **Mass migration.** A herd of saiga antelopes migrates over the vast steppes of Kazakhstan. These nomadic animals never stop moving in search of places to feed and, in spring, to breed.

◀ **Male pose.** A male saiga stands watching for rivals, revealing in profile his horns – sought after by poachers for the Chinese medicine trade – and his proboscis-like nose. This big nose has several functions: dust removal in the dry summer, heat exchange in the cold winter and, in males, nasal roaring in the rut, both to intimidate rivals and to impress females.

## Mass births and mass deaths

The vast steppes that stretch across the middle latitudes of central Asia, from Russia through Kazakhstan and Mongolia to northern China, form the largest continuous expanse of grassland on the planet. But to survive here, you need to be able to cope with major changes in the weather. The temperature can swing from baking to freezing, varying from summer to winter by as much as 80°C. Add the frequent dust storms, and you really do have a kind of hell on earth. No wonder so many wild creatures here are only seasonal visitors.

Yet one creature doesn't just survive here all year round, it dominates the landscape. With its strange nose, the saiga antelope looks like an alien, and in some ways it is. Saigas have been here for tens of thousands of years, once sharing these vast steppes with sabre-toothed tigers and mammoths, both now long extinct. They have done so by adopting one of the oddest of life cycles.

They are able to digest food few other animals bother with, including lichens. Their hooves are shaped to dig through heavy snow, so that in the depths of winter they can locate the vestiges of food that lie beneath. When lakes and rivers freeze, they obtain water from ice and snow. And when chased by wolves, they can flee at speeds of up to 80kph (50mph). But it is that extraordinary nose that is the most striking of the saiga's many adaptations. The swollen protuberance acts as a dust filter, a crucial

**Winter endurance.** Saigas in search of a new area where they can access grass under the snow. Their long, thick, white winter coats are needed on the central Asian steppes, where the temperature can be -35°C (-31°F) or lower. In a severe winter, weakened after the exertions of the midwinter rut, the majority of males may die.

advantage in this windswept landscape, and enables it to keep cool in hot weather and warm in cold weather by recirculating the air, acting as a radiator or air-conditioning system.

Male saigas have a larger protuberance than females, which they use in the rutting season to amplify their bellowing. The larger the nose, the deeper the sound, which may help a male triumph over less well-endowed rivals and mate with a whole harem of females. Even so, a male's chances of survival are low: in some rutting seasons nine out of ten die from their injuries or from sheer exhaustion.

◀ **First legs.** A newborn saiga calf, its twin lying nearby, takes its first steps in the early evening. Calves are born in the night and then left alone – they cannot walk yet, and a mother's presence might draw attention to them. Their mothers pick them up late the next day when they are strong enough to walk and join the herd.

▶ **The big birthing.** Saiga females graze on new growth after rains on a calving ground in Kazakhstan. They all give birth to twins within a week in spring. This minimizes the risk for each calf of being killed by predators such as wolves.

The saiga's unusual adaptations to its harsh environment extend to their calving, which is concentrated into an intense period of just a week or so, when every single female in the herd will give birth. For the waiting predators – wolves, foxes and eagles – this signals a bonanza of easy food (a single wolf can kill six saiga calves in the space of an hour). Such carnage may make mass birthing seem madness, but by synchronizing her calving with her herd-mates, each mother hugely increases the chances that her own baby will survive – a clear demonstration of safety in numbers.

But there are hidden dangers to the saigas' herd lifestyle. From time to time they suffer mass die-offs, caused by an unidentified pathogen or a sudden and severe shortage of food, often in the depths of a particularly severe winter. And then there is human hunting. Whole herds have been gunned down for their horns, to supply the traditional Chinese medicine trade. Now classified as critically endangered, with numbers having plummeted from over 2 million to under 100,000 animals in the past few decades, how long can this incredible survivor manage to hang on in such a hostile world?

▲ (left) **Hiding.** A speckled wood caterpillar, perfectly camouflaged among the grass stems, can feed while hidden.

(right) **Feeding.** A green huntsman spider, relying on camouflage rather than webs, has captured a hoverfly and starts to suck out its juices.

## Resurrection rains

It has now been two weeks since the last rain fell on a hidden corner of Norfolk, in the east of England. But to the west, gathering storm clouds signal a change. The first sign is a drop in the air temperature, followed by tiny raindrops spattering the dusty ground. Minutes later, the skies open. Just a few days later, the meadow starts to transform into a kaleidoscope of colour, with the bright blue of cornflowers, the custard-yellow of corn marigolds and buttercups and the subtle pink of fumitory. Nearby, white ox-eye daisies grow in profusion, alongside swathes of red corn poppies. And throughout, a variety of subtler grasses flower, adding to the diversity and richness of the habitat. It's this ability of grasslands all over the globe to burst into life with the coming of the rains that is one of the key factors that makes them so widespread.

▲ (left) **Roosting.** To avoid predators during the day, a nocturnal grass moth roosts in dense grass.

(right) **Emerging.** Having just emerged from the soil, where it developed as a larva, feeding on plant roots, a cranefly reveals why its popular name is daddy-long-legs.

While the grasses rely on the wind to spread their pollen, each of the flower species has evolved to attract different insects to spread theirs. Following the rain, the insects and other invertebrates emerge in their millions: beetles and bugs, crickets and grasshoppers, flies, butterflies and day-flying moths – a vast hum filling the air as the insects fly low over the meadow in search of life-giving nectar and pollen.

But this is now an increasingly rare sight in much of lowland Europe. The intensification of agriculture and the use of herbicides and pesticides to root out any plant or insect crop 'pests' have led to a massive reduction in the variety and number of these traditional hay-meadows and the disappearance of the pollinators and their food plants. The only hope for many grassland plants is that their seeds can lie dormant in the ground for years – even decades – and one day bloom again when there is a change to the way the land is managed.

▲ **Safety tail.** A harvest mouse climbs a dew-covered reed, using its long prehensile tail as a fifth limb to keep its balance.

◄ **High living.** Resting in its summer nest, woven from shredded leaves high up among reed stems, a harvest mouse investigates a noise. Nests are used for resting, sheltering and giving birth to up to eight tiny young.

## Tree-house living

As the June sun rises over the fields and meadows, and a skylark rises into the air to deliver its reel of song, a movement in the tall grass signals the appearance of one of the world's smallest mammals, the harvest mouse. Its scientific name *Micromys minutus* translates as 'tiny little mouse', which is very appropriate for Europe's smallest rodent. Weighing just a few grams, it is adapted to spend its whole life high up in the grass-forest. If it were any heavier, it would be in danger of bending the grass or even breaking it. No other mammal and only a few birds share this habitat, allowing the harvest mouse to thrive here. The one downside of such a small size is an increased body surface to volume ratio, making it very vulnerable to cold weather, and so it must feed constantly to replenish lost energy.

The harvest mouse's other adaptation – rare among mammals, especially those in the Old World – is its fully prehensile tail. Like the New World monkeys, it uses its tail as a fifth limb, wrapping it around the stems of grass as it moves through the vegetation, aided by its tiny toes and their ultra-sharp claws, which keep a tight grip.

These features have enabled the harvest mouse to colonize a wide range of grassland microhabitats, including cereal crops (especially oats, barley and wheat) and field margins, saltmarsh grasslands and even reedbeds. It is especially active around dawn and dusk, presumably to try to avoid the daylight predators. In spring and summer it can take advantage of the abundant supply of seeds and grains, and in the autumn, fruits, berries and the occasional insect.

Another adaptation is the creation of hanging nests. It uses its ultra-sharp teeth to shred blades of grass, which are then expertly woven into a ball-shaped structure, suspended like a tree house in the canopy. Several nests are woven, used for sleeping, sheltering from bad weather and, of course, breeding.

A litter is usually born in May and comprises about half-a-dozen pea-sized 'pups'. These grow rapidly in their grass-house nursery, and once their eyes open, their parents constantly forage for food for them. The pups start to explore outside the nest when they are 11 days old, and when they are 14–16 days old, they are usually left to their own devices. Their parents may then start another litter and have several before winter arrives.

While foraging, a harvest mouse has to avoid being spotted by predators, including foxes, stoats and weasels, but also the aerial ones, whether generalists including rooks and crows, or specialists such as kestrels and barn owls. To combat them, the harvest mouse has developed extremely acute hearing; as soon as it senses danger, it will freeze or, if necessary, drop down into the grass. But humans present a different kind of danger. In a field, the hay or crop is likely to be cut in late summer, and any youngsters that have not left the nest and found shelter in the field margins will perish in the blades of the mower. Many will also die in a cold, wet winter. Though they build a winter nest for shelter and will sleep for long periods when it is very cold, they are so very small that they find it difficult to keep warm enough.

▶ **Silent strike.** A hunting barn owl uses its wings to slow its flight on approach and then tucks them in to plummet down into the long grass and grab its rodent prey.

## The insect-beaters

The profusion of life on the African savannah is legendary. But though the most visible animals are the great herds of herbivores and the birds that fly above them, the most abundant life here – as in almost every habitat – are insects and other invertebrates. They provide the food for a wide range of small and medium-sized creatures, especially birds.

Of all the bird families adapted to feed on flying insects, the bee-eaters reign supreme. The 25 or more species are found across much of Africa,

**Colourful outriders.** Flying alongside a herd of elephants and a moving film vehicle in Botswana's Savuti Marsh, carmine bee-eaters pick off crickets and other insects flushed out by both feet and wheels. Migrating south, they have stopped off at Savuti after the rains to feast on the vast number of insects that have emerged with the flush of grass.

Europe, Asia and Australia, mostly in warmer climates. But they don't just catch and eat bees. Bee-eaters feed on a wide range of other flying insects, including grasshoppers, ants, beetles and dragonflies, using their long, narrow wings and tail to manoeuvre through the air, twisting and turning in pursuit of their prey.

Botswana's grasslands are home to several species, some of which, such as the southern carmine bee-eater, fly here on their migration south to take advantage of the abundance of food that follows the wet season. At this time of year, the grasslands are at their fullest. So the crickets are safe – as

◀ **Useful elephants.** Darting in among the elephants' feet, the bee-eaters use their long wings to manoeuvre and grab insects.

▼ **Bustard riding.** Using a kori bustard as both a convenient perch and an insect-beater, a carmine bee-eater goes on a leisurely hunt, sometimes reaching over to grab a cricket without even opening its wings. Just occasionally the bustard – also an insect-eater – will get grumpy.

long as they stay hidden in the long grass. But the carmine bee-eaters enlist the help of the biggest animal to roam these plains, indeed the largest of all land mammals, the African elephant. They simply wait until the elephants approach, and then swoop down, weaving among their legs to snatch the panicking crickets out of the air.

Sometimes mobile perches are used for insect-hunting. Carmine bee-eaters will hitch a ride on the back of a range of animals, including the avian equivalent of the elephant, the kori bustard. This species, the world's heaviest flying bird, also feeds on large insects such as grasshoppers and locusts but forages mainly on the ground, though it sometimes, just like the bee-eaters, follows large mammals such as elephants to pick up flushed insects.

▲ **Attired and ready to leap.** A male Jackson's widowbird gets ready for his courtship display with all the concentration of a human high-jumper. He has prepared his court and its grass sculpture, and his breeding plumage is fully grown.

## Grass stage-management

Another spectacular African bird has turned tall grass to its advantage, using it in an astonishing courtship display. Jackson's widowbird is found in upland grasslands in central and western Kenya and northeast Tanzania, though much habitat is now under threat from the intensification of agriculture. It is the only member of its family to engage in courtship at a communal 'lek', where males perform against each other to impress the watching females. Lekking is a 'winner takes all' strategy, in which the triumphant male gets to mate with the majority of the available females.

On the high plains, after the rains have begun and as the flowering grasses reach their full height, the male widowbirds gather in grassland arenas. Each male prunes a circle around a central tussock, where he will perform his display. Throwing back his head, ruffling out his neck feathers and arching his long tail over his back, he leaps up to a metre (3 feet) in the air. But often the surrounding grass is so tall that the females, perched

▲ **Practising the high jump.** A trial leap. Only when he knows females are watching from nearby in the grass does he leap high into the air, brandishing his long, glossy black tail. Females will also judge him on his gardening and topiary skills.

on the grass stems, simply can't see the males. Some clever males have turned this to their advantage. Using the long grass as a stage curtain, the performer reveals himself momentarily. It is almost as if, by rationing his appearances before the female, he makes them even more special. But he must be careful: if he shows himself too clearly, he risks being grabbed by an opportunistic predator such as a serval.

If his trick is successful, he waits for the female to land nearby and then dances around her, fluffing up his crown and neck feathers. Only if the female is really impressed will she mate with him – and for the vast majority of males, this will never happen.

As with all lekking species, a male Jackson's widowbird takes no part in raising the young, though he will alarm-call if predators are nearby. Unfortunately, the biggest problem faced by these ground-nesting birds is having their eggs or chicks trampled by feeding cattle. There is also a race for the females to raise their family before the seasonal glut of food finally runs out.

## The listening cat

The rainy season ended months ago, and the unrelenting sun has beaten down on the African savannah ever since, turning what was briefly a lush, green oasis into a shimmering scene of yellow and brown. For the big cats – the lions, leopards and cheetahs – the change is to their advantage. Their prey animals are now concentrated at or near waterholes, which makes them vulnerable. The colour of the grass helps the big cats, too: their sandy-brown coats allow them to blend in and stalk their prey unseen.

Also living on the savannah are small cats: the African wildcat, the caracal and the serval. They target small mammals, along with birds, frogs, insects and even the occasional snake. Rodents are the main prey of the serval. These can be hard to find in the wet season, but as the grass begins to dry out, they become easier to see and hear. Two adaptations make the serval an effective hunter. The first is its sheer athleticism. Servals are able to jump more than 2 metres (almost 7 feet) skywards and as much as 6 metres (20 feet) along the ground, using legs that are longer in relation to their body than those of any other cat.

But rats and mice can also leap and put on a turn of speed. So the serval has developed huge ears that locate and pinpoint the exact range and direction of prey. Even so, patience is key. The process begins with the serval walking very slowly through the long grass, stopping every few minutes to listen. Sometimes it may stay put for as long as 15 minutes, waiting to detect the tiniest rustle that indicates a rodent. Using its spring-loaded legs, it pounces, striking the prey with one or both its paws. If the rodent evades the initial strike, the serval may try again or bounce around the area on stiff, straight legs, which may even flush out a bonus meal.

Only when the winds blow strongly across the grassland does the serval not bother to hunt, for then it is difficult to hear. But in the right conditions, the combination of camouflage, stealth, acute hearing and athleticism make the serval a very successful grassland predator.

◀ **The pounce.** Having swivelled forward its ears to pinpoint the rustle of a rodent, a serval moves into spring position ready to pounce on it with both forepaws. The serval is the tallest of Africa's small cats, and has incredibly long legs – perfect for hunting in long grass. Its large, funnel-like ears help it locate prey.

▲ **The art of cutting.** Chaco leaf-cutter ants saw off sections of fresh grass, cutting them width-wise with their serrated mandibles. The grass is carried back to the nest for the underground fungus garden, which provides the food for the growing colony.

## The army of grass-cutters

On the other side of the world, in Argentina, in the vast Gran Chaco savannah, it is now late in the dry season, and the larger herbivores have moved on to pastures new. Only stiff blades of sword grass remain, and most of the grasses' nutrients have been stored underground in readiness for the next rains, when a grassy forest will emerge.

For the vast majority of creatures, there is little energy to be gained from the meagre vegetation left above ground. But the leaf-cutter ants have evolved a way to obtain the vital nutrients they need from what remains. They harvest it and feed it to a fungus, which they farm underground in their nests. The relationship with the fungus is mutually beneficial. The ants keep it free of parasites and provide it with food in the form of grass and leaf material, on which the fungus grows. In return, the fungus provides a constantly harvestable food source for the whole colony, including the millions of ant larvae and their queen.

In fact, leaf-cutter ants consume far more vegetation than any other group of animals and are the dominant herbivore of the neotropical region. With up to 7 million workers in a colony, each one able to carry more than 50 times its own body weight, an average nest could well use more

▲ **The transport column.** A column of Chaco leaf-cutter ants carries grass segments several times longer and heavier than themselves back to the nest. The smaller workers are marking the trail and guarding against parasitic flies attacking the labouring workers.

than 180 million separate pieces of grass – a phenomenal task requiring extraordinary levels of cooperation and organisation.

Leaf-cutter ants build enormous nests, the bulk of which lies underground. These structures, along with those of termites, are among the biggest constructed by any animal. The army of workers it houses are divided into castes, performing a variety of roles: excavation and construction, garbage disposal, foraging, gardening, chopping up the leaves and grasses, distributing the chopped material to the fungus, tending the fungus and nursing the queen and her larvae.

Protecting the colony, inside and out, are soldiers. But perhaps the greatest threat to the colony is rain. During the wet season, the nests are regularly inundated, preventing the ants from foraging. The ants stop the water getting into the nest by blocking any openings with tiny twigs or fragments of clay. They cannot deal so easily with another threat: parasitic phorid flies. These tiny flies target the larger worker ants. Their attack is rapid: a female fly lands on a worker ant's head, then inserts its ovipositor and lays a single egg. This will hatch into a larva, feeding off and eventually killing the ant. But the inroads the flies make on ant numbers are minimal. The ant colonies remain key to the productivity of the grassland, recycling plant material and, in turn, providing food for a range of wild creatures.

## Biggest herds and greatest fortresses

The world's grasslands are home to an astonishing variety of creatures, not least the huge herds of large grazing mammals that roam these vast open spaces. Yet even these multitudes are dwarfed by the billions of individuals that make up 'herds' of termites. The world's 3000 or so species of termites are not related to ants but share an ancestor with cockroaches. What they do have in common with ants is that they live as 'superorganisms', and they also consume grass – more than all the other creatures on the planet put together.

Rather than grazing on new growth, they exploit dead and dying grass. Like all insects, their perennial problem is that they are an abundant food resource for many other animals. Their defence is to build huge mounds, protecting themselves against attack from predators and sudden changes in the environment, as well as providing food-storage space for when things get tough. These fortresses stand citadel-like on grasslands across the tropics and have enabled termites to colonize every continent apart from Antarctica and cope with major changes in seasonal weather. In a state of constant flux as they adjust to the surrounding world, the mounds seem almost alive, maintaining the same internal temperature as mammals.

Australia's Northern Territory is home to a particularly unusual species of termite. The magnetic termite builds its mounds in islands of low-lying grassland surrounded by savannah woodland. These are flooded for part of the year when the monsoonal rains arrive, so the termites cannot retreat underground and must survive the summer heat in the mound itself. The mounds, which are up to 4 metres (13 feet) high, are always oriented in a north–south direction, which maximizes the termites' ability to maintain their home at a constant temperature and allows the rapid circulation of air to keep the inner chambers well ventilated. When the sun is at its hottest, at midday, just the edge of the mound faces the sun. When the sun rises, the western face of the mound is shaded, and more termites are to be found in galleries on this side, and in the afternoon and evening, when the eastern side remains shaded, more termites move to this side.

In the cooler nights, the workers venture outside to harvest dead grass, which they chew into pellets and then store in the mound's chambers. They depend on them when the rains come and they are trapped inside. Aboriginal Australians hold these termite mounds in high esteem, believing

▲ **Solar positioning.** In Australia's Northern Territory, magnetic termites align their huge wedge-shaped mounds from north to south, to minimize the heat during the hottest part of the day. The large surface area of the flat shape also helps the mounds to 'breathe'.

them to be the resting sites of their dead ancestors. But they have also learned a practical lesson from these tiny insects: to orient their homes in a north–south direction to keep the temperature as low as possible in this unforgiving cauldron of heat. More recently, architects elsewhere have studied the termite mounds in order to create low-energy buildings.

But termite mounds are not impenetrable. As with ant colonies, the abundance of potential food attracts predators, including some peculiar animals – the aardvark, for example. This mostly nocturnal mammal, with

▲ **Ant mining.** On Brazil's Pantanal grassland, a giant anteater uses its powerful claws to dig up an ants' nest. It then sucks up the ants, aided by its long, sticky tongue. Its thick skin and long hair protect it from any bites or stings.

◀ **Termite tunnelling.** At night, in Botswana, an aardvark digs deep into a termite mound, licking up the termites with its long tongue. It is being attacked by soldier termites but has a tough skin that helps protect it from stings and bites.

powerful claws and a long, pig-like snout (its name means 'earth pig' in Afrikaans), is the only living representative of an entire order of animals. It lives in equatorial and tropical Africa, where it roams over grasslands, smelling out its favourite food, termites. It will then dig rapidly into a mound, using its sharp, powerful claws to penetrate the outer casing, and its long, sticky tongue to vacuum up thousands of termites at a time. A cohort of robin-like chats often follows in its wake to pick up any termites that it leaves.

Several other animals have evolved to exploit the abundant, if tricky to access, ants and termites. On the savannah and open grasslands of South America lives the giant anteater, capable of attacking an ants' nest with such destructive force that it has been compared with King Kong. It destroys a nest with a single blow of a powerful clawed foot and then, like the aardvark, uses its long tongue to feed. But the ants are far from defenceless: swarms of soldiers emerge to attack an invader, and like all social insects, they are prepared to die for the good of their relatives inside the nest.

## Solving the snow problem

▲ **Benefiting from bison.** In winter in Yellowstone National Park, Wyoming, a red fox watches for rodents flushed out by the bison as they shovel away the snow.

► **Snowploughing.** In the bleak winter landscape of Yellowstone, bison survive by digging down to the grass, excavating up to 6 tonnes of snow a day in the process.

Far from the searing heat of the Australian and African plains, another kind of grassland is also home to vast numbers of animals. The great swathes of prairie that stretch from Alaska and Canada through the north of Europe to the eastern tip of Asia may look bleak, but a number of animals have evolved to take advantage of the seasonally abundant supplies of food to be found there.

Among these is one of North America's best known mammals, the bison. Millions once roamed these northern plains, but with the coming of the European settlers, the vast majority were killed, and only a tiny population remains. And like all animals this far north, they are vulnerable when winter comes and snow covers the grass. If more than a metre of snow falls, a young bison is in big trouble. To help it, a mother will use her powerful head as a snowplough so her offspring can get to the life-giving food beneath.

The red foxes that live alongside the herds of bison face a similar challenge. But in their case, the food is the rodents such as mice and voles that live on grass. Under the snow out of sight of any predators, these little animals continue to feed on grass, but the foxes listen out for them. Tapping into the Earth's magnetic field, a fox will align itself to the northeast and nose-dive into the deep snow. Using this technique, it has a high hit-rate,

often pouncing on the right spot and emerging from the snow with a vole between its jaws. So far, the fox is the only mammal known to use the Earth's magnetic field to gauge both the direction and the distance to its prey.

Farther north, on the Arctic tundra of northern Canada, another drama is being played out. Covered with snow for more than half the year, these vast, exposed grasslands present different challenges. But one animal has adapted to thrive there in vast numbers: the caribou, or reindeer, as it is known in Europe and Asia. Everything about caribou, from their double-layered, thick fur to their specially padded hooves that act as snowshoes, help them cope with the extremes of climate. Even the food they eat – a lichen known inaccurately as reindeer moss – is only digestible in their special stomachs.

Caribou can run fast, but they are always in danger from wolves, especially when the calves are born. The females' plight is heightened by the fact that, before calving, they separate from the main herd and head north to their calving ground where there is nutritious food and fewer biting insects but without protection from the larger males.

Caribou give birth to a single calf in early June, when the toughest winter weather should be over, though there is always the risk of a late snowfall or even an ice storm, which has been known to wipe out a whole generation of calves. As with saiga antelopes, the births are synchronized to minimize the risk of predation for each individual. Up to 50,000 calves may share the same birthday. They are soon able to stand on their spindly legs and suckle, but in a matter of days they have to be strong enough to begin the long trek back to the rest of the herd.

If the herd is attacked on this epic march southwards, the females can outrun the pack, but usually at least one calf is brought down and killed. The majority, though, evade the attack and continue on to rendezvous with the main herd. Like their mothers, they have learned the secret of survival on these vast northern grasslands: stick together and never stop moving.

◀ **Snow baby.** A mother reindeer (known as caribou in North America) encourages her calf, just hours old, to stand and walk so that she and it can move with the other females and calves to rejoin the main herd on Russia's Taimyr Peninsula.

▶ (next page) **Birthing ground.** Female reindeer graze at a birthing ground in the Taimyr region of central Siberia – part of the largest population of migrating wild reindeer in Eurasia, 700,000 strong. Always on the move to find new grazing, these are among the most nomadic of animals, roaming in huge herds.

# 5
# ISLANDS

**SURROUNDED BY SEA, ISLANDS ARE**, by their very nature, worlds unto themselves – home to some of the most specialized and spectacular wild creatures on the planet. Islands are as varied as they are numerous. They range in size from Greenland – almost nine times larger than the United Kingdom – to tiny, wave-splashed rocks. Whether they are desert islands, coral atolls or covered with snow and ice, they all have one thing in common: isolation.

Islands form natural boundaries, limiting arrivals from outside but also making departure difficult. Birds may be able to come and go, but most other animals have no choice but to make the best of their surroundings. Many species found on islands are the descendants of castaways, whose ancestors swam, drifted or floated there millennia ago. Having been isolated for so long, they have evolved into unique forms, ultimately becoming different species from their mainland relatives.

By doing something different, an island animal is able to thrive and may even become ruler of its tiny kingdom. So paradoxically, though an island may support fewer species than the adjacent mainland, those found there are often unique – sometimes to a region or an archipelago, but often to a single island. Thus though islands comprise less than one sixth of the Earth's land area, they are home to more than 20 per cent of its bird, plant and reptile species.

In the absence of their old predators or competitors, many animals (and plants) have developed unusual characteristics. Some are bigger than their mainland cousins – the Galápagos tortoises, for example. Others, such as

(previous page) **Penguin heaven.** The volcanic island of Zavodovski, a remote Antarctic haven for chinstraps in the South Sandwich Islands, where 1.5 million gather to breed, safe from most predators and surrounded by fish-rich seas that provide the food for their chicks.

◀ **Island refuge.** The coral-fringed Isla Escudo de Veraguas, off Panama, the only home for several species, including the Escudo hummingbird, the Escudo fruit-eating bat and the pygmy sloth, which lives in the mangrove forest on the shores of the island and its islets.

the tiny Inaccessible Island rail, are far smaller. Flightlessness among birds is also far more common on islands than on the mainland, simply because, in the absence of natural predators, they don't need to get airborne. But for many species, such evolutionary developments have ultimately proved fatal.

When human explorers finally reached the remote islands of the Indian and Pacific Oceans in the sixteenth and seventeenth centuries, they brought with them rats, cats and dogs. These soon made short work of helpless island birds such as the dodo, whose plump body and inability to escape also endeared it to hungry sailors desperate for fresh meat. But today, there is greater awareness of the precariousness of island life. Scientists are working hard to stop more species going extinct on these 'living laboratories', where we can study the unique behaviour of the wildlife that makes its home there.

► **Island pygmy.** On the tiny island of Escudo de Veraguas, a pygmy three-toed sloth, with her baby attached, climbs a red mangrove in search of leaves to eat. She is much smaller than three-toed sloths on mainland Central America, a result of her species having been isolated for 9000 years, living on a restricted diet. Her fur is coloured green by algae, which provide camouflage and additional snacks.

## The smallest sloth

For millennia, wild creatures have become island castaways, either drifting across the ocean until they wash up on land or becoming stranded when sea levels rise and turn their mainland home into an island. Among them are some of the most peculiar animals on Earth.

Off the mainland of Panama, in the Caribbean Sea, lies the Isla Escudo de Veraguas. The size of New York's Central Park, the island is relatively new, having become isolated from the Central American mainland less than 9000 years ago. When the seas rose, several species were cut off from their mainland cousins, and in a surprisingly short time evolved into separate species. One of these was the brown-throated three-toed sloth. It may have been a good swimmer, but it wasn't good enough to swim the 17km (10.6 miles) back to the mainland. In its isolation, it shrank by about 40 per cent, weighing in at less than a domestic cat. It also grew longer hairs on the crown and sides of its head, giving it a hooded appearance, and its skull changed shape. Today scientists consider it to be a full species, the pygmy three-toed sloth.

Like all sloths, it leads an arboreal (tree-climbing) lifestyle, appearing to spend most of its life in the same patch of red mangroves, where it feeds on leaves. Compared with the diet of mainland sloths, these provide relatively few nutrients, which may go some way to explaining why these island animals are so much smaller than their relatives.

◀ (top) **Swimming action.** A pygmy sloth crosses between patches of mangrove at high tide. The quickest way to cross the mudflats is to wait for the tide and swim, and pygmy sloths are good swimmers.

◀ (bottom) **Looking for love.** At mating time, a male pygmy sloth crosses to another mangrove beach. If a female is on heat, she will howl to attract potential mates, and swimming is often the quickest way to reach her.

Pygmy sloths are excellent examples of 'insular dwarfism', when a species grows significantly smaller from generation to generation, until it stabilizes at a new size and weight. There are several explanations for this process. The most plausible is that, when a population becomes separated and trapped in one place and a food shortage occurs, the smaller animals get by on less food and pass on their genetic predisposition for small size to the next generation.

Insular dwarfism doesn't just happen on oceanic islands: it can also occur in species found in other isolated habitats, such as caves, oases and mountaintops – often known as 'sky islands' because their ecology mimics that of real islands. But it happens most often at oceanic locations.

On Escudo de Veraguas itself, the pygmy sloth is confined mainly to mangroves, which now cover not more than 10 hectares (25 acres). With such a restricted range and small population – there may be as few as a couple of hundred – it comes as no surprise to discover that the pygmy sloth is officially listed as critically endangered. Yet for those that do remain, it's not a bad life: all the food they can eat (if the mangroves remain intact). Being smaller, they need to feed less often than their mainland cousins, so can sleep for even longer. Indeed the pygmy sloth's lifestyle could almost be described as idyllic, were it not for the fact that it is imprisoned on its island paradise.

## The biggest tortoise

Small may not always be the best route to survival on an island, and many creatures display the opposite trait: they become giants. Nowhere is this more apparent than on the Galápagos Islands. Off the Pacific coast of South America, straddling the equator, the Galápagos gained their worldwide fame as the cradle of evolutionary theory when Charles Darwin visited in the 1830s. Observing the strange life forms he encountered, adapted to different situations, led him to develop his theory of evolution through natural selection.

Of all the creatures he came across, few were as impressive as the giant tortoises. They weigh on average roughly the same as three grown men. Once, these huge reptiles roamed across many of the world's continents, but today they are only found on two island groups, separated by thousands of miles of sea: Aldabra in the Indian Ocean, and the Galápagos, where they live on 7 of the 21 islands. These reptiles are not only huge but also among the longest-lived of any animal: surviving more than a century in the wild (up to 170 years in captivity). Of the 15 known races, 11 survive today.

Like many of the other creatures on the Galápagos, the giant tortoises were castaways, drifting across the ocean from mainland South America. They managed to travel such a distance – some 1000km (620 miles) – because they float and can survive for several months without food or water. Helped by ocean currents, though some would have drifted westwards, enough made landfall for a viable population to become established.

Darwin was stunned by the size of these reptiles, noting in his journal that it took up to eight men to lift a tortoise. So why are they so big? One theory suggests that only the largest tortoises could have survived such a difficult sea crossing, as their large volume to surface area ratio would have allowed them to retain fluids, while their longer necks meant they could breathe more easily.

The other notable feature is how the shape of their shells differs. The shapes seem to correlate with the habitat of the island on which they live: those on lush, well-vegetated islands have domed shells and shorter necks; those on dry, desert-like islands have 'saddleback' shells and long necks. Presumably, where there is plenty of low-growing vegetation, the animals can easily reach their food, while on drier islands, they may need to reach up to get to taller vegetation such as the abundant prickly pear cactus. However, other scientists have suggested that this may be a coincidence and that the saddleback shell shape is actually a product of sexual selection, a result of generations of females preferring males with a higher front to their shell and longer necks. But then two races that have a similar shape may not necessarily be very closely related, which tends to suggest that they have evolved in response to the type of habitat in which they live.

Famously, at the time of his visit, Darwin failed to appreciate the differences in shell-shape between each island tortoise. Only when he returned to England did he develop his theory; and then, to his eternal regret, he found that he had not always labelled the tortoise shells he had collected with the name of the island where it had been found – thus making the specimens far less useful for determining whether his assumptions were correct.

When the islands were discovered in the sixteenth century there were as many as a quarter of a million giant tortoises; today there are just a few thousand. But with better habitat conservation, and the removal of introduced animals, their future now looks more assured.

▶ **The biggest survivor.** An Aldabra giant tortoise, on the remote Indian Ocean atoll of Aldabra – part of the Seychelles group. It is more than a metre long (more than 3 feet), and is the only species of giant tortoise to have survived in this part of the world, the others having been hunted to extinction by visiting seafarers or settlers or their habitat destroyed by introduced animals accompanying humans. These giants can live for more than 100 years, probably more than 150.

## Here be dragons

If you want to see a true giant, travel to a handful of more than 17,000 islands that together make up the archipelago of Indonesia. Here you can find the largest and most fearsome lizard on Earth: the Komodo dragon. First described little more than a century ago, in 1912, the species may have been known to ancient cartographers, who marked maps in this area with the ominous warning 'Here Be Dragons'.

This huge reptile is well named: growing to 3 metres long (9.8 feet) and weighing up to the same as a large adult male human. It has had more than 4 million years to evolve, ever since the island was first cut off by rising seas. It was also very well adapted as a castaway: tough and strong, with a leathery skin. And with no larger competitors to worry about, it has become the undisputed ruler of its island kingdom. Like almost all lizards, the Komodo dragon must bask in the rays of the tropical sun each morning before it has enough energy to be fully active. Even then, it rarely looks as if it is in a hurry, lumbering slowly and deliberately across the island terrain, in search of a potential meal.

It has an acute sense of smell to catch the scent of any passing victim, and often ambushes its prey, hiding surprisingly well for such a large creature, and then showing an equally surprising turn of speed. Though it takes a wide variety of creatures – birds, mammals and even invertebrates – it is not averse to feeding on carrion: a dead deer does not require as much energy to deal with as a live one and is just as valuable a food. And not being fussy is an advantage on an island where you have to take what you can get. The Komodo dragon also has a very effective weapon. When it bites its prey, it secretes venom that, though it may not kill the animal immediately, renders it incapable of living more than a few hours afterwards. The dragon then waits patiently for its victim to die – most victims die of blood loss before the venom has time to take effect – or become weak enough to pose no threat.

It is unusual for the top predator in any environment to be a reptile. But living on an island means the dragon has few competitors. It has also adapted its breeding behaviour: female dragons are capable of parthenogenesis – they can give birth to males without their eggs being

▶ **On the scent of prey.** Scenting prey with its tongue, a Komodo dragon races across the beach, drooling saliva full of bacteria that will help kill any animal it bites.

fertilized by a male – though they are still able to reproduce sexually. This gives them an added advantage, allowing colonization of an area by a lone female mating with her sons.

They don't always hunt as individuals; uniquely among lizards, Komodo dragons also come together in packs to maximize their chances of a kill. Communal feeding can lead to squabbles between individuals, but normally the smaller animal will give way. Komodo dragons occasionally attack humans, with a single bite from those terrible jaws sometimes proving fatal; but they are more likely to take the easier option of raiding shallow graves.

Comparing these gigantic reptiles with their considerably smaller counterparts on the Australian mainland, it would be easy to assume that their size is a result of the process of 'gigantism' that creates so many large herbivores on oceanic islands. Yet though we think of the Komodo dragon as a giant, in fact it appears to be an example of a process by which large animals isolated for long periods on islands become not larger, but smaller. It seems that Komodo dragons may actually have evolved from

**Grappling giants.** Two rival male Komodo dragons – the world's largest lizards – wrestle each other on Rinca Island, one of the islands comprising Komodo National Park. The bout lasted just seconds before one was knocked over onto his back.

an even bigger – and now long extinct – mainland relative, which at up to 7 metres (23 feet) long would have been more than twice the length of the dragons we see today.

Evolutionary biologist Jared Diamond believes that those gigantic prehistoric lizards evolved to hunt another extinct creature, the dwarf elephant *Stegodon*. This survived until just a few thousand years ago on the nearby island of Flores, where almost half the 4000 or so Komodo dragons still live today. The fact that there are two closely related species of smaller dragons on both Flores and Komodo would lend support to the idea that, rather than evolving from a large reptile that became even bigger once it had taken up island life, the opposite is the case. This supports the view that as the home range of a large animal decreases in size, so do the animals themselves.

It seems that all these species, including the Komodo dragon we see today, may have evolved from even larger creatures, and so are in fact – like the pygmy three-toed sloth on the other side of the world – examples of insular dwarfism, rather than gigantism.

## Spaghetti-like snakes

As the pygmy three-toed sloth has shown, there are some advantages to becoming a dwarf, such as being able to cope better with food shortages, the ability to survive on smaller prey items and, especially, the need to reproduce more rapidly in response to sudden and extreme changes in their surroundings. But surely, if you are a predator, the advantages of being bigger and stronger must outweigh the disadvantages? Yet in the forests of the Caribbean island of Martinique, there lives a predator so tiny and so easy to miss that it was only described for science as recently as 2008.

The Martinique threadsnake is just 10cm (4 inches) long and as thin as a strand of spaghetti, and it rivals its cousin the Barbados threadsnake as the world's smallest snake. Given that the two other smallest snakes in the world are also found on isolated Caribbean islands (Barbados and St Lucia), there must be some advantage to being so tiny. In this case it is because there are few other predatory creatures on these islands, allowing these tiny snakes to squeeze into small spaces and exploit the abundant food resource of ant and termite eggs and larvae. But there is a cost to becoming so small. Whereas most snakes lay up to 100 eggs, a threadsnake lays only one; were it to lay any more, its hatchlings would be too small to survive. This makes its offspring very vulnerable to predation, having foregone the usual 'safety in numbers' strategy of most reptiles. It also means that the species multiplies very slowly, making it vulnerable to external factors such as loss of its forest habitat or sudden and extreme weather events such as hurricanes.

Nevertheless, the dwarfism route is one that many island forms of reptile and amphibian have gone down: the scientist who discovered the Barbados threadsnake has also found one of the world's smallest frogs, on a mountain 'island' on Cuba, and the world's tiniest lizard, the Jaragua gecko, on an isolated, predator-free island in the Caribbean. So long as the niche you fill is unoccupied, it makes sound evolutionary sense to be either tiny or, if you are a tortoise on the Galápagos or Aldabra, very large indeed.

▶ **Thread-thin snake.** The world's smallest snake burrows through leaf-litter in forest on the Caribbean island of Martinique. This threadsnake – the smallest a snake can be – is able to squeeze into tiny spaces in its search for termites and ants. But the cost of such a narrow body is that it can only produce one egg at a time. Its tiny mouth means that it sucks up termite and ant eggs like sweets, and can't dislocate its jaws as other snakes can.

▲ (left) **Aye-aye** – the endangered, nocturnal and rarely seen Madagascar lemur, revealing its elongated middle finger, used for tapping wood to locate cavities where insect larvae might be lurking and for hooking them out. Its huge ears indicate that it listens for its prey, though its diet is mainly seeds, fruit and nuts.

(right) **Madame Berthe's mouse lemur** – the world's smallest-known primate. Restricted to a small area of dry forest in Madagascar, it feeds at night, mainly on fruits and gum, and sugary insect secretions.

## The many shapes of lemurs

Island life doesn't just lead to extremes in size – from gigantic to tiny – it also allows life-forms isolated from the mainland for millions of years to evolve into unique shapes, too. And the place to go for shape-shifters is one massive island: Madagascar. This is the oldest extant island on Earth – having been isolated from the African mainland for at least 165 million years, and finally breaking away from the Indian subcontinent about 88 million years ago. There is no better place to look for evolutionary oddities. Madagascar truly is a living laboratory – indeed its fauna and flora are so different from any elsewhere in the world that it has been dubbed 'the eighth continent'.

The island is almost as large as Alaska and more than twice the size of the United Kingdom. It boasts no fewer than 10,000 species of plants (more than 80 per cent of its plant species) that occur nowhere else; also unique are 96 per cent of the island's reptiles, 60 per cent of its birds, 100 species of fish,

▲ (left) **Black-and-white ruffed lemur** – the largest of the lemurs (along with the red-ruffed lemur) on Madagascar and critically endangered. It eats mainly fruit, seeds and leaves, and also drinks nectar, pollinating the flowers of at least one tree in the process, making it possibly the world's largest pollinator.

(right) **Golden-crowned sifaka** – another fruit and leaf eater, confined to forest patches in just one region of Madagascar and like so many other lemurs, critically endangered, mainly through loss of its habitat.

all 650 or more species of land snail, and most of the island's butterflies. But it is Madagascar's unique mammals that are truly fascinating.

Of the island's 200 or so living species, more than one third belong to a single group: the lemurs. They have evolved to fill every kind of tree-dwelling niche. Lemurs are primitive primates, most closely related to those other large-eyed, nocturnal and arboreal primates, pottos and lorises. The lemurs' ancestors appear to have arrived on Madagascar roughly 60 million years ago. They probably came by sea, travelling on mats of floating vegetation pushed almost 500km (310 miles) eastwards from mainland Africa by oceanic currents. They found what must have seemed like an island paradise: no major large mammal predators and, with few other mammal species, a range of vacant ecological niches ready to exploit.

Over time, this allowed the lemurs' ancestors to evolve into very different and distinctive body-shapes and sizes, each to fit a particular lifestyle. Some, such as the ruffed lemurs, evolved to leap through trees like monkeys; others, such as the ring-tailed lemurs, developed a shape that enables them to walk along the ground. With these different niches came very different kinds of behaviour: ring-tailed lemurs are famously social animals, while the mysterious indri – the world's largest living lemur, tipping the scales at up to 9kg (19.8 pounds) – is more solitary.

Their diets also reflect their lifestyles. Some eat insects, others just fruit and nuts, and the bamboo lemurs, as their name suggests, live entirely on rapid-growing bamboo. A number forage for food under cover of darkness, which explains why several species of lemur have only been discovered in the past decade or so.

Two very different species have taken to this nocturnal lifestyle. The large aye-aye moves slowly through the forest canopy, seeking insects and their larvae, which it extracts from holes in the wood using its elongated middle finger. Madame Berthe's mouse lemur – the smallest primate on the planet – weighs just 31 grams, hardly more than an ounce. Confined to one small area of rainforest along the west coast of Madagascar, it feeds on fruit and the sugary secretions of insects.

Verreaux's sifaka has exploited a habitat so harsh that few other creatures survive there. A population lives in spiny-thicket forest in a desert-like area in the south, and these lemurs have mastered the art of leaping from one spiny tree-trunk to another without impaling themselves. A mother must carry her single youngster on her belly or back while undertaking leaps of up to 10 metres (33 feet). When Verreaux's sifakas do descend to the ground, they use a strange gait, hopping and dancing sideways with their arms raised above their heads.

Lemurs came to dominate the mammal fauna of Madagascar through lack of competition and by becoming specialized to fit the vacant niches. For 60 million years or so, this strategy worked well; but in the past few hundred years, the lemurs' specialism has made them uniquely vulnerable. Since the arrival of humans less than 2000 years ago, at least 17 species have become extinct. All were larger than any species that survive today – presumably they were both easier to catch and better to eat. Today, most species of lemur (94 per cent, according to the International Union for Conservation of Nature) are threatened with extinction because of deforestation, hunting and capture for the pet trade, combined with political instability that makes long-term conservation difficult to achieve.

▶ **The power leap.** Using its powerful back legs, a Verreaux's sifaka leaps between branches in spiny-thicket forest in the southern part of the island. Compared to other big lemurs, its arms are small, but its leaps are large.

▶ (next page) **The bounce move.** Crossing open ground between pockets of trees, a Verreaux's sifaka spring-bounds in a unique sideways movement, using her arms for balance. Like so many lemurs, this species is endangered and confined to patches of spiny forest.

◀ **Meeting and greeting.** A female Buller's albatross, possibly 24 years old, greets her mate after returning from half a year at sea. He arrived ahead of her at their nest, built in a contorted woodland of giant daisies on one of the Snares Islands. She has used the nest for more than ten years. Made of mud and vegetation, it keeps the eggs and chicks clear of running water when it rains, which it does a lot on the Snares.

## Sailors home from the sea

On a remote, wind-lashed island in the Southern Ocean, a reunion is about to take place. The couple haven't seen one another for almost half a year, during which time each of them has flown thousands of miles around the oceans of the southern hemisphere, never once making landfall.

They are Buller's albatrosses. The male spent the austral winter in the strip of ocean that runs south from Tasmania; the female travelled all the way to the coasts of Chile and Peru, where she fed in the rich waters of the Humboldt Current. This might explain why she is late for their rendezvous, here on the Snares Islands off the southern tip of New Zealand. He waits anxiously for her to return. Of course she may not come back – many thousands of albatrosses fall victim each year to the industrial longline fishing, snared on the long lines of baited hooks.

If she does make it back, these large seabirds have no time to lose. The summer here is productive but short, giving just enough time for the pair to mate and for the female to lay her single egg, which will hatch out into one of the most comical-looking youngsters in the bird world.

The high seas may seem like a hostile habitat, but to albatrosses it's home; coming to land is far more risky. That is why Buller's albatrosses – almost 9000 pairs of them – choose to breed on small rocky islands in the Southern Ocean, where they know they are safe from predators, for no non-human land mammal has ever made it this far. This is where they return to breed every year – a pair ringed in 1948 was still breeding at the same nest-site in 1971, 23 years later.

Albatrosses are not the only birds taking advantage of a safe place to breed. Sooty shearwaters nest here in their millions, their burrows riddling the muddy woodland floor under the tree daisies. Also known as 'muttonbirds', their tasty flesh was coveted by sailors, who would harvest the plump youngsters. There are also Snares crested penguins – 25,000 pairs of which breed here and nowhere else. More than 3 million seabirds, squeezed into an area of just 3.5 square kilometres (1.35 square miles), shape the island: the shearwaters carry leaves into their burrows, helping to recycle nutrients, while their guano fertilizes the soil.

Finally, the female albatross arrives back, signalling to her mate with a loud, deep call. Now the race begins: will they manage to raise their chick before the autumn gales sweep across this exposed land? The egg

is incubated for more than ten weeks, and the chick takes another 20 weeks to fledge.

While sitting on her nest, the female has ample time to contemplate the nature of her island surroundings. The rocky landscape is dotted with giant herbs: the Macquarie Island cabbage, a giant carrot and a stingless (fortunately) giant nettle, and woodlands of tree daisy, growing to more than twice the height of a human. They have become giants in the absence of native mammal herbivores to graze on them.

If all goes well for the chick, it will fledge and leave the nest six months after its parents landed. But having finally taken flight, it will not reach sexual maturity until it is at least ten to twelve years old. In the meantime it will roam the southern oceans, gliding effortlessly past icebergs until it, too, is ready to return to the place where it was born and to raise its own young.

▲ **Albatross island haven.** Buller's albatrosses on their pedestal nests in a thicket of tree daisies and among giant cabbage (not truly a cabbage but used by past sailors as a remedy for scurvy).

▶ **Little penguins, giant daisies.** Snares crested penguins make their way up a stream-bed to their nests in the interior of the tree-daisy woodland on North East Island, one of the Snares Islands – the only islands where they breed.

## The carnivorous island

As soon as the autumn has passed, the subantarctic islands such as Snares become too hostile to sustain a mass of birdlife. To find a paradise that birds can enjoy all year round, we need to travel to the tropics. The Seychelles islands lie just south of the equator, in the western Indian Ocean; this really is a tropical paradise. Or is it? For despite the endless sunshine, white sandy beaches and vivid blue seas, something dark and sinister is going on. It involves a very strange plant, the pisonia, and a member of the tern family, the lesser noddy.

This elegant seabird nests in vast numbers on the island of Cousin – up to 80,000 pairs squeeze onto this tiny piece of land during the southeast monsoon season from June to August. They build their nests out of leaves and seaweed – but it is where they choose to breed that is particularly unusual. Noddies nest in pisonia trees, where their chicks are safe from the crabs and skinks scuttling across the ground below, at least for a while. For as the chick grows, so, too, do the hooked, barbed and sticky seeds of the pisonia. The noddies are sensible enough to nest in male trees, but the seeds of the female trees fall to the ground, and as the chicks leave the nest and begin to explore the surroundings of their nest, the seeds stick firmly to their growing feathers.

If only a few seeds attach, the noddy can cope, and when it does finally fledge and fly far away, the seeds travel with it, enabling the plant to disperse over a wide area. But this clever trick often backfires: the chick can become so weighed down that it never fledges properly and dies on the ground. No wonder the locals refer to the pisonia as 'the bird-catcher tree'. And presumably, the tree makes use of the handy fertilizer provided by the corpses.

A cousin of the lesser noddy, the fairy tern, avoids the perils of the pisonia by nesting on a branch of the tree. It does not bother to make a nest at all, laying a single egg in a shallow depression on a bare branch. This does keep the egg and chick relatively safe from crabs, rats and lizards, which seldom travel along the narrow branches, and it may also enable the terns to avoid the attention of nest parasites such as ticks and fleas. But it is still a risky strategy: unseasonal winds may dislodge the egg. When this happens, as it frequently does, the tern simply lays another egg. After all, unlike the albatrosses on the Snares, which have to race to complete nesting before the summer is over, the fairy tern has an endless breeding season.

▶ (top) **Pisonia problems.** A juvenile brown noddy reveals its coating of seeds from the 'bird-catcher tree'. As the young bird has walked around, enough seeds have stuck to it to prevent it ever being able to fly. Preening won't remove them.

▶ (bottom) **The sticky end.** The body of a young lesser noddy that has become entangled in a mass of pisonia seeds and starved to death. Pisonia trees fruit twice a year, coinciding first with the arrival of the adult noddies and then with the fledging of the young, giving the tree two chances of dispersing its seeds.

## Birth of an island

Islands are, by their isolated nature, dynamic. In most cases, isolation happened slowly, as tiny shifts in the Earth's tectonic plates or a gradual rise in sea level cut them off from the mainland to which they were once attached. But some islands have a far more explosive birth – emerging from the depths of the ocean in a tiny fraction of geological time. Over the course of a little more than six months, from November 1963 to June 1964, a completely new island emerged off the southern tip of Iceland. Named Surtsey, after a giant from Icelandic mythology, it was the result of an undersea volcanic eruption.

By the time the eruption was finally over, Surtsey had grown to almost 3 square kilometres (1.15 square miles) in area, and though it has gradually diminished under the onslaught of the sea, it still reaches a maximum height of more than 150 metres (492 feet) above sea level.

For scientists – especially biologists – the sudden appearance of this island presented a wonderful opportunity to study how living organisms colonize new land. From the very start they logged the plants and animals that arrived on the pristine black soil: insects and birds that had flown there, plants whose seeds had been blown by the wind, and marine life around its shores. Each new arrival marked a change to the clean slate of this new land.

Today, more than half a century after the island rose so dramatically, species recorded on Surtsey include more than 70 plants, nesting seabirds – among them gulls, fulmars and puffins – and hundreds of invertebrates, many of them flightless, which have hitch-hiked to the island on the feathers of birds or drifted there across the sea.

Surtsey isn't the only new island to have appeared over the past few decades. They have arisen near Japan, off Tonga in the South Pacific and in the Hawaiian Islands, themselves formed by volcanic activity over millions of years. Hawaii is one of the remotest archipelagos, and until humans arrived roughly 1700 years ago, the typical gap between each new species colonizing the islands was 10,000 to 100,000 years. All that has changed. Humans attract hitch-hikers, and today, Hawaii gains 20 new species every year, which are wreaking havoc, driving many of its native species to extinction.

◀ **Newborn island.** Surtsey, the product of a volcanic eruption off Iceland less than 60 years ago. It is now colonized by plants and animals that have blown, flown or hitch-hiked there.

## Water lizards

If you want to be successful as an island colonizer, it pays to be at the head of the queue. Any plant or animal that gets there first has the chance to dominate the whole ecosystem or at least adapt to its new environment before competitors or predators arrive. On the island of Fernandina – the youngest yet also the largest of all the Galápagos islands – one reptile has become king by arriving soon after the volcanic island emerged from the sea.

Since this iguana arrived on a piece of driftwood or matted vegetation from mainland South America, about 8 million years ago, it has adapted to a whole new way of life. As its name suggests, the marine iguana has taken to life in the sea, making it unique among the world's 6000 species of lizard.

Marine iguanas spend much of the time lazing about on the hard, black rocks around the island's shoreline. Nothing much disturbs them – adult iguanas have no predators (though the babies do) and few, if any, competitors. When it's hungry, it simply slides off its rock and swims out to sea, where it dives down to forage for food, particular algae (seaweed), which can also be found along the tideline. The bigger the iguana, the deeper the dive.

But having evolved as a land animal, it struggles to maintain its body temperature in the cold waters around the island, and the smaller iguanas can only do one long dive a day, the timing depending on the times of the tides and when they can warm up. Afterwards, a marine iguana hauls itself back onshore and warms its body in the equatorial sun, while the oddly named Sally Lightfoot crabs pick skin from its body, and lizards catch flies around it, making a living from what is the dominant animal here.

So life for the iguanas is pretty easy – until, that is, they breed. Like many reptiles, they lay their eggs in shallow burrows – in their case, on the sandy beach. But in June, when the babies emerge, predators are waiting. Galápagos hawks and frigatebirds pounce on the youngsters when they emerge. But this aerial bombardment pales into insignificance when compared with the hidden danger from the land: Galápagos racer snakes. Enough survive, though, to keep them island kings.

Despite Charles Darwin's view of them as 'disgusting, clumsy lizards … imps of darkness', even he was unable to ignore the extraordinary ways in which this entrepreneurial lizard has adapted to its island home, exploiting an ecological niche in a way no other member of its tribe has been able to do.

► **Island sun worshippers.** A pair of marine iguanas, with young iguanas using them as they would rocks, warm up after sunrise on a beach on the Galapágos island of Fernandina. This lizard is a true entrepreneur colonizer, having adapted to feeding in the sea, on offshore marine algae.

► (next page) **Diving for dinner.** A marine iguana grazes on seaweed in shallow water off Fernandina. The water is cold, and though iguanas can hold their breath for quite a time, after a while they have to return to shore to warm up. Larger iguanas don't lose heat as fast and so can stay active for longer and dive deeper.

## Race for life

Marine iguana eggs are eaten by many of the introduced animals on the Galápagos, and even by some of the native birds. And when the babies hatch and emerge from the sand, they run a gauntlet of snakes as they head to the shore. Galápagos racer snakes know when the bonanza is due and wait for it. If a baby walks slowly, it may not get noticed, as snakes see movement better than detail. But as soon as it runs, they will give chase, sometimes as many as 20–30 of them together. On the flat, it can outrun them, but if it heads for the rocky lava, it enters the killing fields, to be caught by a writhing ball of snakes and swallowed alive.

**1 The watchers.** Galapágos racer snakes wait for the hatching of the baby iguanas. They hole up in the lava rocks, often in groups of ten or more, watching for movement.

**2 The chase.** A baby iguana makes the mistake of moving fast. The racers spot it and give chase down the beach.

**3 The scrabble.** Having wrapped themselves around the iguana, the racers scrabble to get their mouths around the head and claim the meal.

**4 The end.** Several snakes keep hold of the back end of the baby but will eventually give way to the snake that has now managed to engulf the head. It will swallow the iguana alive.

## Red crabs and crazy ants

▲ **On the move.** A female Christmas Island red crab, holding her fertilized eggs under her, makes her way to the seashore. She has just emerged from mating in a male's burrow. Normally sedate, at migration time, the crabs develop metabolic superpowers.

► **Back from the sea.** Female crabs return from the sea. The tide was right and their eggs were ready to hatch and were released into the sea. After a month, the babies will return to the island to become land crabs.

If you live on an island, you can never become complacent, even if you are a marine iguana. For if a single factor changes, you may find yourself trapped and unable to escape from your fragile paradise. Christmas Island lies in the Indian Ocean, well off the northwest coast of Australia, though part of it is geographically and ecologically much closer to New Guinea. As with so many oceanic islands, its fauna and flora are fairly limited, a factor that allows those species that do live there to thrive and prosper, in some cases in their millions. Of these, the best known are the various species of land crab. These land crabs have made the opposite journey from the marine iguana's

▲ **Crazy ants.** High on sugar, yellow crazy ants swarm over a red crab. They are killing it, not to eat but because the migrating crab has crossed into their territory. These introduced ants now form supercolonies on Christmas Island and are decimating the crabs and, as a result, the island ecology – the crabs are vital forest gardeners.

one, leaving the sea for life on dry land. But they still need to enter the sea to spawn, as their embryos must develop in the sea. So every November the island's beaches play host to one of the greatest wildlife spectacles, as up to 80 million red crabs head towards the sea in an epic and potentially fatal journey. Birds wait in ambush, as do robber crabs. But the biggest threat is a far smaller and yet far more dangerous creature.

The yellow crazy ant got its name from the jerky movements it makes when disturbed. Its origins are unknown, though it may have arrived on produce from Southeast Asia. Today, it can be found throughout the tropics, and has been described as one of the worst invasive species in the world. At low densities, it is not a problem, but when they have access to energy sources, the yellow crazy ant can dominate any native ants. It sets up supercolonies covering vast areas. It also attacks far larger creatures,

▲ **Party booty.** A party of yellow crazy ants carry away a piece of red crab claw – a taste of protein. Their normal diet is honeydew secreted by scale insects, but having overpowered the crab by squirting acid into its eyes and then its joints, here they are making use of part of it, carrying it back to the supercolony.

including the red crabs, that enter their territories, but also when the crabs migrate to the beach. And they do so in a number of grisly ways: baby crabs may be eaten, but mainly the ants attack any crabs, squirting acid in their eyes and blinding them, purely as defence rather than to eat them. Crab numbers have halved, and if this decline continues, the species could disappear over the next few decades. This would have a devastating effect on the island: red crabs dig burrows that aerate the soil and feed on weeds that otherwise would change the nature of the forest. Currently, the authorities are trying to control the ants by controlling the scale insects that secrete the honeydew the ants feed on.

If the red crab does disappear, it won't be the first island creature to be dominant at one moment and go extinct the next: 80 per cent of all known species extinctions since 1500 have taken place on islands.

## Turning hell into paradise

As the Christmas Island red crab illustrates, however well adapted you may be, when the world changes, your strengths soon become weaknesses. So is there such a thing as an island utopia?

Zavodovski Island is a volcanic subantarctic island in the South Atlantic, part of the South Sandwich Islands chain. It's a forbidding place – but not if you are a chinstrap penguin. Zavodovski hosts the world's largest penguin nesting colony, with more than 1.3 million pairs. The reason so many penguins come to islands such as Zavodovski is that they are free from land-based predators. Even so, penguins must keep a lookout for giant petrels, which are always circling on the lookout for unguarded eggs or chicks. Zavodovski's shoreline is also battered by waves, which throw the penguins around like rag dolls, as they try to land or go back to sea. The gravel slopes also emit sulphurous fumes from the island's volcanic centre, but for the most part, the wind is strong enough to blow the fumes away.

A pair of chinstraps is faithful to a particular spot, returning every November to the same little patch, and also to one another – most keep the same partner. A pair will take it in turns to head out to catch fish to regurgitate for their two chicks, as one has to remain on guard against marauding skuas. The cacophony of tens of thousands of penguins is disorienting for humans, but chinstraps, like other colonial-nesting seabirds, have the ability to pick out their mates' calls and those of their chicks above the constant din.

It is the abundance of food in the surrounding seas that makes this seemingly hellish place a true heaven for the chinstrap penguins. For ultimately, whether you are on an island or any other place on the planet, life is all about having what you need to survive and reproduce. And with well over a million penguins migrating to Zavodovski each year to breed, this island and the sea really must be paradise.

◄ **Landing waves.** Hundreds of chinstrap penguins try to land on a tiny, steep stretch of exposed beach, battling against waves driven by 30-knot winds. They must surf ashore on the incoming waves and then run like crazy. This is risky: the churning sea means that many penguins get stuck between boulders and are injured or killed.

► (next page) **Volcanic vista.** A view over a small part of the chinstrap penguin colony on the north side of Zavodovski, in January, when most of the pairs have young chicks.

# 6
# CITIES

**THE CITY NEVER STOPS**. Here the natural cycle of the day is broken, and urban dwellers live in a 24/7 culture. Today, more than half the world's human population – almost 4 billion people – live in towns and cities. And by the end of the century, more people could be living here than are alive today – roughly 9 billion.

This is a very recent phenomenon. There are trees alive today that are about the same age as the oldest city on the planet, and just 200 years ago only 1 in 30 people lived in urban areas. So you might expect that wildlife – with life cycles governed by tens of thousands, or even millions, of years of evolution – would struggle to adapt to city life. But the world's cities provide food, water, shelter, places to raise a family and countless opportunities and rewards for species and individuals able to adapt to a different way of life.

The urban environment is constantly changing, with new dangers but also opportunities. Being able to alter your habits and behaviour to fit the new landscape is crucial. From woodland birds nesting in artificial boxes and owls setting up winter roosts, to hyenas as garbage cleaners and leopards as street prowlers, many creatures have adapted to city life in one way or another.

An important characteristic of cities, from an animal's point of view, is their diversity, offering a range of mini-habitats. These fall into two categories. The first is what ecologists call 'encapsulated countryside': relict fragments of wood and meadow, lakes and rivers, and the occasional patch of heath or moor, that were present before people arrived in numbers and began to

(previous page) **City cat.** A leopard drinks from a well overlooking Mumbai, India, one of the world's most crowded cities. While the people sleep, the leopards that live in this metropolitan park slip into the streets to hunt pigs, dogs and other feral animals.

◀ **Singapore's super-trees.** Designed to catch rainwater, hung with native plants and fitted with solar panels, these 50-metre-high (164-foot) structures light up at night. They are part of a plan to put green space and green architecture in the centre of Singapore.

urbanize the land. The second includes all the new habitats, for example, the parks, gardens, cemeteries, rubbish dumps, canal banks and roadside verges. We didn't create these as habitats for nature, yet they are often some of the very best places for wild plants and animals to make their home.

The result of this fusion of patches of habitat – natural, semi-natural and entirely man-made – is that the world's cities are home to almost any group of plants and animals you can imagine. You can find large and small mammals, marine creatures, birds and insects, reptiles and amphibians, wild flowers, grasses and trees, and many small creatures, sometimes at far greater population densities than in the wider environment.

Of course, life in the urban jungle is not always easy for wild species. For a start, they have to deal with humans. If they fail to adapt, they will die. But for those that do have the ingenuity, resilience and staying power to survive, the rewards can be great. It's all about being streetwise.

▶ **Freerunning langur-style.** A female langur, baby attached, clears a huge gap between rooftops, her tail curled for balance. Regular routes used by the langurs of Jodhpur involve spectacular leaps, often made at speed. Because people provide food for them on the rooftops, some groups of langurs in this urban canopy may never even come down to street-level. At night they sleep in the trees.

## The holy primates

Few wild animals are quite so successful in an urban environment as the Hanuman langur, one of Asia's largest monkeys. Named after the monkey god who led a troop of monkeys that helped rescue Sita, the god Rama's wife, from a demon king – a battle of good versus evil – it plays a special role in the Hindu culture of India. Its black hands and face are symbolic of the burns Hanuman suffered in his heroic battle.

It's not an animal you'd expect to thrive in the heart of a city. Indeed most of India's 300,000 or so langurs live in dry tropical deciduous forest and spend at least half their lives in trees. They are well adapted to the arboreal lifestyle, able to leap almost 5 metres (16 feet) from branch to branch, and to drop more than 12 metres (40 feet) to the ground.

Yet in the city of Jodhpur, in northwestern India, a population of Hanuman langurs leads a very different lifestyle. Troops of these city monkeys, each led by an alpha male, use the city buildings and structures as they would trees and rocks, leaping between walls and rooftops. But instead of eating natural vegetation, they rely on food provided by humans.

A lead male has his work cut out in this urban jungle. There are gangs of young males hanging around looking for any opportunity to unseat him, and as in rural areas, his tenure is seldom more than two and a

◀ **Monkey playpen.** A group of young Hanuman langurs play in the safety of a courtyard, practising leaping and jumping. The blue-painted city of Jodhpur offers plenty of play areas and freedom from predators.

▶ **Feast day.** It's Tuesday – Hanuman day – and worshippers at a Hindu temple in Jodhpur have put out extra food for the Hanuman langurs. People also regularly provide food on the rooftops, often when they prepare their own meals, so that at least one troop of langurs has adapted to a human schedule of eating three meals a day.

half years. But though city life is more stressful, he is likely to sire more young, and that's because the females benefit from city life.

A female begins breeding at about three-and-a-half-years old – about half the age of rural langurs, and can give birth every year afterwards, whereas most langurs do so only every 18–24 months. She is helped by the fact that other females in the troop assist with care of her youngster. A mature female can live as long as 35 years, during which time she and her offspring will have brought up several generations of langurs – possibly double the number she could have produced in the countryside. And in a country where drought can be a problem, Jodhpur also provides access to water.

The reason urban Hanuman langurs do so well is because of the easy availability of food. They are mainly the beneficiaries of city people, receiving almost all their food from people. They don't need to beg or steal. People not only tolerate their presence but encourage it, putting out food for them, sharing their picnics and letting them feed in their gardens, all in honour of Hanuman. Langurs are simply beloved. At Hindu shrines, food is laid out for them, and in the Hindu temples, it has been a longstanding tradition to feed them every day, but especially on Tuesdays. If a langur dies on that holy day, it will be given a funeral.

As with other holy animals such as the cows that roam many an Indian road or street, Hanuman langurs are also protected by law from being captured or killed. In no other country are animals treated as well and are regarded as having the right to live alongside humans.

## The clean-up clans

When it comes to large, fierce creatures, few are tolerated by the people they live alongside, and most are actively persecuted or even exterminated, and they fear humans as much as we fear them. So why would the spotted hyena, one of Africa's most formidable predators, choose to enter the ancient Ethiopian city of Harar?

Here in this hilltop settlement, where a labyrinth of narrow streets and mud buildings surrounded by high stone walls creates a traffic-free network more than 400 years old, these fearsome hunters thrive. During the heat of the day, the hyenas usually sleep in dens outside the city walls.

▲ **Clan stand-off**. On the outskirts of Harar, members of one hyena clan stand moaning and groaning, neck hairs bristling, as they stare intimidatingly at the place where another clan has gathered. It is a stand-off at an invisible line, probably marked with scent. There are three clans, and when they move into the city at night, they coexist, seemingly through the establishment of dominance hierarchies, with few skirmishes.

But as dusk falls, they head towards Harar, entering through special hyena gateways in the ancient walls and into the heart of the city.

They have come in search of food, especially the bones put out by the butchers. With their incredibly strong jaws they crunch them up, providing a service for the city. But what is truly unique is the hyenas' relationship with this Muslim society.

Elsewhere in Africa, hyenas take livestock and occasionally small children. But here they never kill animals or harm people. It is as if they have a pact with the population, one that has evolved over hundreds of years. Certainly, there is one family that has been feeding them over six generations and has a relationship with these animals that is so strong

◄ **Hyena gate.** A hyena makes its way to a hyena gate in Harar's city wall. Hundreds of years ago, these entrances were built in the city walls, large enough for hyenas to enter but too small for a besieging army.

► **Show time.** To prove the intimacy of his relationship, Yusuf uses his mouth to give Willi II meat. He is a fifth-generation hyena man and knows the names of every hyena in the clan that lives near to his house, providing them most nights with scraps from the butchers and the tannery. Though Willi II is not a high-ranking hyena (females are the clan leaders), he is bold and so often gets to be fed first.

► (next page) **Going to the meat market.** On the way to scavenge in the heart of the old city, hyenas walk nervously past barking dogs in the police station. The hyenas are not so nervous of people, though, and the people take little notice of them. In the market square, the butchers have left out bones and meat refuse for the hyenas to clear up – a practice that has gone on for centuries.

that they can even feed them mouth-to-mouth. They also recognize the individuals and have given them names.

By helping the three clans of hyenas that regularly visit the city, the people of Harar protect themselves against attacks by other clans or wandering individuals. That clans coexist mainly without fights seems to be because dominance struggles usually occur at designated meeting places rather than at boundary lines. Even though these encounters look terrifying, few hyenas get wounded, as the encounters have become ritualized. This may explain why, when they are in the city, these hyenas are relatively tolerant of one another – they have already worked out the hierarchy and know each other by smell as well as visual cues. Also, within the city walls there is no grass and the streets are cobbled and so boundary marking may not be so effective.

The other service the hyenas provide for the people of this highly religious city is to gobble up bad spirits – jins. Elsewhere in Africa, the hyenas' famous cackling sound is often regarded with fear and linked to witchcraft; but here it is thought to ward off evil spirits, and so is welcomed – just like the hyenas themselves.

◀ **Break-in.** A raccoon attempts to lever open the window of an empty house. Studies in the USA appear to show that urban raccoons are quicker to solve problems than country-living ones, probably because the more adaptable raccoons are the ones that survive in urban situations.

▶ **Handiness.** Having learnt to steal eggs from an open hen house by day, a raccoon attempts to extract an egg from it at night, using its dextrous and sensitive hands.

## The raiders

It's hard to imagine people in European and North American cities being quite so forgiving of animals, especially considering our attitude towards gull raiders or dumpster-diving raccoons. But that hasn't stopped these clever creatures from continuing to live in cities. In some North American cities, raccoons have learned to open doors with their dextrous hands and squeeze through the smallest of gaps.

The toughest time for any creature is when it first ventures into the urban environment. So after the kits leave the safety of their den, the mother raccoon guides them through the cityscape, teaching them the techniques they will need to survive. Like hyenas, raccoons keep to the shadows, usually foraging only under the cover of darkness. Their insatiable curiosity enables them to find the plentiful food discarded by our wasteful society – and may also get them into trouble – raccoons are notoriously mischievous and often end up in unlikely spots. But it may be that having to solve the many problems thrown up by urban living has actually made this wily creature even more intelligent. Another reason for the success not just of

▲ **Market raiders.** Part of a huge troop of macaques retreats to the safety of a rooftop to eat the fruit they have stolen from the market. When they raid as a group, there is not much humans can do to stop them.

the raccoon but also of other medium-sized mammals such as the fox, is that they are of just the right size to succeed in cities. Few large predators such as big cats are likely to find enough food in urban settings, and most will be killed or driven out before they have a chance to settle down. But medium-sized animals such as foxes and raccoons are the right size to make best use of human facilities.

On the other side of the planet, another urban animal is far bolder. In the northern Indian city of Jaipur, 3.5 million people have to contend with

one of the cheekiest animals of all, the rhesus macaque. It lives in troops of dozens of individuals, and like the Hanuman langur, is associated with the Hindu god Hanuman and is often found around temples and other sacred sites.

But unlike the langurs, rhesus macaques can be real pests, mainly because of their boldness, which includes stealing food, especially from the open displays in city markets, and entering houses. What is remarkable is that they are still venerated, even in more cosmopolitan

parts of India. In the capital, Delhi, where the macaque population is estimated to be in six figures, these curious monkeys often go where they shouldn't, causing damage and inconvenience.

They also frequently put both themselves and their human neighbours in danger by using electric power cables as they would branches. In this major city, the tolerance of such mischievous creatures to the extent of allowing them to share many of the same urban facilities – far more so than the high-living langurs – is extraordinary. Solutions to perceived problem animals have included playing loud sounds, using langurs to frighten off the smaller macaques and even forced sterilization. But despite the conflicts, the people of Delhi continue to tolerate the presence of these charismatic fellow primates, not getting rid of them but regarding them as neighbours with a right to remain as urban dwellers.

▲ **City squatters.** From the window of their home, two macaques watch the activity below. The human occupants have left, and now the macaque troop takes its midday siestas and spends the night in the house.

▶ **Macaque mischief.** Ever playful, young macaques wrestle, learning adult social skills in the process.

## The red toy thief

The great bowerbird of northern Australia could be said to be interested in the urban potential for interior design. Indeed, it spends much of its time being creative. It lives in a range of habitats, from mangrove swamps to tropical forests. But like many other Australian birds, it has also learned to live alongside humans – in particular, in Townsville, Queensland.

The male of this pigeon-sized bowerbird builds an elaborate bower – a twig-lined, 1-metre-long (3-foot) avenue, leading to a court, on which he performs to any interested female who will watch from the avenue. To make himself even more appealing and entice her to mate with him, he decorates his court with what at first sight looks like a random assortment of rubbish but is, in fact, a carefully chosen, colour-coordinated arrangement of ornaments to appeal to his potential mate.

Different individuals have a preference for different-coloured objects, often prizing the rarer colours, especially red. In their natural habitat, males make do with crimson berries and scarlet flowers, but here in the city they can do much better than that. As a result, Townsville's children learn not to leave any small red toys out on the garden lawn.

But if a male does manage to find a red toy, he doesn't just bring it back and plonk it down anywhere. He spends hours meticulously rearranging the bower's contents to enhance the display and impress the female with his diligence. He even arranges objects by size, small to large, creating an optical illusion known as forced perspective, perhaps to make himself look bigger but certainly to make the display more attention-grabbing.

If his design skills are approved of, the female great bowerbird mates with him and then leaves; he takes no part in raising the offspring. She does all the nest-building, incubation and chick care, while he just carries on maintaining and improving his bower.

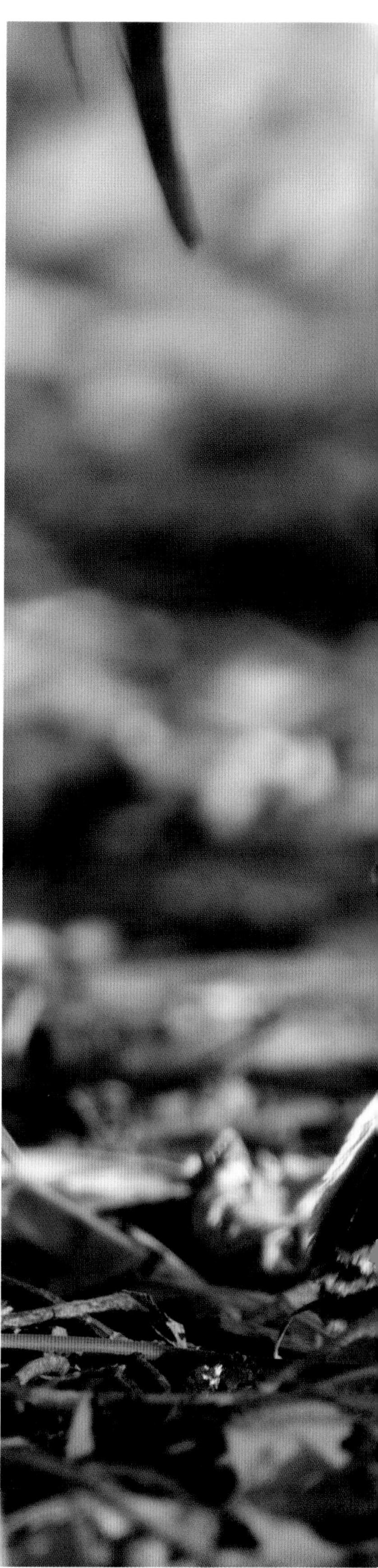

▶ **The art of recycling.** A male great bowerbird moves a red object away from his arrangement of white objects and to the outskirts of his bower, where he likes the reds to be. He knows they will be a particular attraction to females. In Townsville, Australia, the bowerbirds favour plastic objects, presumably because they are not perishable and can be reused year after year.

◀ **Showing off.** A male great bowerbird inspects the view from his bower, taking no notice of the traffic on the Townsville road behind. Laid out on the court are a carefully arranged selection of green glass and white and grey plastic and stones. The red objects are carefully placed around the edge.

Bowerbirds are known to live up to 30 years – longer than any other songbird. Through experience, older male great bowerbirds have a clever strategy. Instead of searching around for objects to work within their colour scheme, they simply check out the bower of a nearby younger male and steal whatever takes their fancy. It usually works; females tend to be more impressed by older males, perhaps because they have learned exactly what works in the mating game.

Occasionally, though, the tables are turned. A resident male mistakes a young male approaching his bower for a potential mate (the plumage is similar to a female's). While he concentrates on his elaborate – but this time pointless – courtship display, the youngster steals one of his precious ornaments and takes it back to his own bower.

## Mass movements to the city

Food and sex are two of the most important things in any wild creature's life. A third one is safety. And for smaller birds in particular, staying alive means seeking safety in numbers.

Gathering in huge flocks to roost has several advantages. First, it makes it harder for a predator to pick off an individual. Second, especially in cold weather, huddling allows each bird to stay warmer than it would be on its own. And finally, communal roosting may allow birds to find food the next day by following the fittest, healthiest-looking individuals to wherever they are feeding.

Not all birds roost at night. Nocturnal species such as owls seek quiet spots during the day, where they can avoid the unwelcome mobbing of songbirds. Single owls are more vulnerable to mobbing than larger groups, which is one reason why, in the town of Kikinda, in northern Serbia, hundreds of long-eared owls gather each winter to roost.

They could roost in the surrounding countryside, but coming into an urban area provides advantages. The presence of humans helps to deter predators. And a town or city will be several degrees warmer than rural areas all year round. This is because buildings act as storage heaters, retaining the sun's heat and then radiating it out after dark, and vehicle emissions also help raise the temperature. Known as the 'urban heat-island effect', this can make a huge difference to city

wildlife, enabling plants and animals to survive the winter when rural temperatures would be too low for them. But the Kikinda owl roost is a relatively recent phenomenon, having been present for just 15 years. The first few birds must have attracted others, and today more than 700 long-eared owls roost in the small town square – by far the largest gathering of this species anywhere in the world. Being so easy to see, the owls have now become a tourist attraction and a source of great pride to the local people.

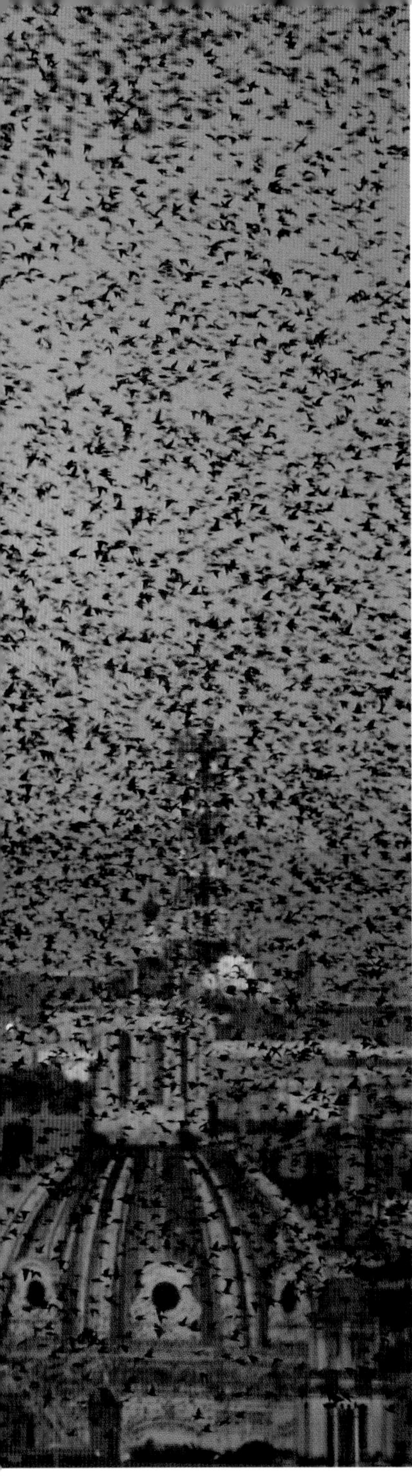

▲ **Winter pilgrims to Rome.** On a December evening, a huge flock of starlings descends on the centre of Rome. Several million of the birds – most of them migrants from northern Europe – use the city in winter for night-time warmth and safety from predators.

About 1000km (622 miles) to the southwest, in the Italian capital of Rome, is a far greater gathering of birds, several million. The locals are not so welcoming of the massive flocks of starlings that twist and turn over the city skyline on winter evenings. Like the owls, the starlings have come to this urban centre for warmth and safety.

Where starlings roost in large numbers in Britain – for example on Brighton or Aberystwyth piers or on the Somerset Levels – they have become tourist attractions. But there is far more for the visitor to Rome to enjoy besides this starling spectacle. So the response to the starling roost has mainly been negative, not least because these swooping birds leave their mark on the city's buildings and the cars in the form of millions of sticky white droppings, and the streets have to be cleaned of droppings daily. Also, the night-time noise beneath the trees where they roost can be deafening.

One creature, however, is delighted that the starlings are here. The peregrine is the world's fastest animal, capable of reaching speeds of more than 322kph (200mph) in its hunting dive, or 'stoop'. In Rome, it takes advantage of this abundant and reliable source of food. But with so many targets to choose from, hunting peregrines have a quarter less chance of a catch than if they were pursuing a solitary individual. Even so, they keep trying, perhaps because the starlings present the falcons with some kind of super-stimulus. They struggle to make a choice, and sometimes they end up hungry.

## Hunting from the city cliffs

Across the Atlantic in New York, peregrines are having a much easier time. In fact they are doing very well indeed, partly because, unlike most urban wildlife, they have hardly had to adapt at all to city life.

Peregrines are in one sense specialist hunters, killing during an aerial pursuit, stooping down towards their intended target at terrifying speed before grabbing it with their ultra-sharp talons. But in another way they are generalists, able to kill almost any small or medium-sized bird. So as long as there are birds around, they have plenty to eat. And thanks to our habit of feeding garden (or in North America, backyard) birds, there are an awful lot of birds in our cities.

Having caught its prey, the peregrine will carry it to a perch nearby, where it will pluck it, then either eat the bird itself or take it back to its chicks, somewhere on top of a tall skyscraper nearby.

In doing so, it has not had to change its normal behaviour in the slightest, unlike most of the other creatures in this chapter. For the tall building on which it nests is the urban equivalent of a cliff-face or high mountain crag, the space in which it hunts is simply the place where birds fly, and its prey is what it has always eaten: birds. So though for most urban wildlife the transition from rural to city living is often not easy, for lucky creatures such as the peregrine, the built environment almost exactly mimics their ancestral home, creating what scientists call an 'analogue habitat'.

Fifty years ago the story was very different for the peregrine, and indeed for many other birds of prey right across the northern hemisphere. The widespread use of chemical pesticides such as DDT in intensive agriculture devastated raptor populations by thinning their eggshells, which massively reduced breeding success. In the nick of time the cause of the problem was discovered, the chemicals were banned, and peregrines began their comeback, thanks also to reintroductions into the wild in New York State.

In 1983, two pairs of the hand-reared reintroduced peregrines turned up unexpectedly in the heart of New York City, liked what they saw, and began to breed. Since then peregrines have started to colonize cities on both sides of the Atlantic, and the birds are now regularly nesting on tall buildings in many cities in North America and Europe. In New York City, they now breed at higher densities than anywhere else in the world.

But the most successful example of a bird that has swapped a wild habitat for city life is the pigeon. That's partly because they have been living alongside us for so long: the rock dove, the wild ancestor of city pigeons all over the world, was first domesticated at least 5000 and possibly as long as 10,000 years ago.

Its success has revealed a surprising phenomenon. So comfortable are pigeons in the urban environment, with such an easy abundance of food and, despite the arrival of the peregrine, so few predators and competitors, it appears that their evolution is now 'going into reverse'. Normally, weak or sickly pigeons die and so are removed from the gene pool. But in many cities, where they can find food, they survive and reproduce, thus passing on their poor genes to successive generations.

▶ **High-rise view.** A peregrine falcon powers off its high-rise roost in New York City. This high vantage point gives it the perfect view of its favourite prey, street pigeons. And as the city is at the hub of the Atlantic-flyway bird-migration route, peregrines can take their pick of any decent-sized migrating bird. Such ideal conditions for hunting and 'clifftop' nesting mean that New York now has the highest density of nesting peregrines anywhere in the world.

## Catfishing for birds

In a stretch of river flowing through the city of Albi in southwest France lives a population of introduced catfish – now Europe's largest freshwater fish – that have learned to hunt feral pigeons. No fishing is allowed here, and so the catfish have grown large enough to power up the bank and suck in pigeons whole, but catfish bigger than 2 metres (6.6 feet) risk stranding if they attempt a snatch. The pigeons have also learned that the catfish are there and are very nervous about drinking along this stretch of river. Once they start bathing, though, they appear to forget. The splashing and the oil from their feathers are what attract the catfish.

**1 The stalk.** A medium-sized catfish stalks pigeons drinking at the edge of the city river. It is one of a group of individuals (identifiable by the pattern of their spots) that have learned to hunt pigeons.

**2 The lunge.** Sneaking up on a bathing pigeon, having used its whisker-like barbels to sense the movement, the catfish lunges at it.

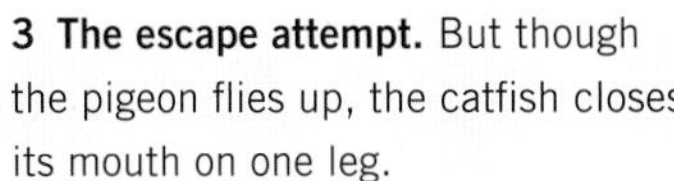

**3 The escape attempt.** But though the pigeon flies up, the catfish closes its mouth on one leg.

**4 The end.** Held by a leg, the pigeon is pulled under water and is then swallowed whole.

## The many moons

From the birth of the first cities, more than 5000 years ago, until the middle of Queen Victoria's reign, the only artificial light to illuminate the darkness was fire. Then, in the late 1870s, all that changed virtually overnight with the invention of the incandescent light bulb. Little more than a century later, the world has been transformed. Many parts of the globe are now illuminated 24 hours a day, to the extent that cities can easily be seen from outer space. As a result, light pollution has become a major issue – not just for people, but for wildlife too.

▲ **Shanghai nightlight.** The lights of central Shanghai, China – one of the world's most populated cities – a famous evening display, which can even be seen from space. But like the lights of other city skyscrapers, they can be a deadly lure for the birds that migrate along the coast. Migrating shorebirds are also affected by the reclamation of Shanghai's coastal wetlands.

When a moth takes to the air, it navigates by using the moon: so long as it keeps the moon at the same angle to its body, it can fly in a straight line in the direction it wishes to go. But when that light is just a few metres away, it all starts to go wrong. If the moth succeeds in keeping the light parallel to its course of flight, it will inevitably begin turning, getting closer and closer to the source of light until eventually it reaches it.

Naturalists take advantage of this by using a mercury vapour light bulb, with emissions similar to that of the moon, to lure moths into a trap. But they aren't the only ones to use artificial light as a lure. Bats gather around streetlights where they know there will be large congregations of moths.

Some spiders deliberately build their webs near streetlights – again, because it increases their chances of catching more insects.

Geckos, too, catch flying insects around lamps, indoors, using their ability to climb vertical walls and walk upside down across ceilings to get to their prey. The geckos' adhesive feet evolved to enable them to hang onto slippery tropical leaves, but city geckos can climb walls and windows just as easily.

In urban areas, true darkness is now a thing of the past. Ecological light pollution – as this phenomenon is known – has many unintended results. Songbirds such as European robins usually sing from just before dawn and again at dusk; both are times when the air is often still, so sound carries further, and when it is generally too dark to forage for food. Normally they then feed during the day and sleep at night. But in cities such as London, the illumination of streetlights throughout the night fools some robins into thinking the sun is about to come up, so they start to sing. Neighbouring males then join in, to ensure that they do not lose out to their rivals in territorial or mating stakes. But this 'arms race', in which all robins in a well-lit urban environment end up singing through the night instead of sleeping, is now having a negative effect on their circadian rhythm.

Birds that migrate long distances fly by night, choosing to travel under cover of darkness for several reasons: the air is cooler, which prevents them overheating, and they are more likely to avoid predators. But the key reason for nocturnal migration is that songbirds use the moon and stars, along with the Earth's magnetic field, to navigate. Like the moths drawn to the light, when their flight-paths take them over cities they become disorientated and can no longer find their way.

We have long known that lighthouses attract migrating birds, many of which end up flying into the light and perishing long before they ever reach their destination. But recent evidence suggests that millions of small birds are dying each spring and autumn when they crash into well-lit buildings in our cities. In North America, city authorities are

▶ **Death by light.** A school group examines some of the more than 1000 birds, representing 89 species collected over 3 months, which have flown into skyscrapers in central Toronto, Canada (another 1000 survived the collisions and were released). Turning off the lights of high-rise buildings would save birds and money and reduce pollution.

◀ **Lure of the nightlight.** On Juno Beach, Florida, a leatherback sea turtle hauls up to lay her eggs. When the baby turtles hatch, they head for the brightest horizon, which should be the natural glow of moon on the sea. But light from beachfront development can lead them off course, and wandering hatchlings can be hit by cars or caught by predators.

▶ **Moth trap.** In Hong Kong, a tokay gecko hangs out on a street light, which provides an ideal moth lure. Some species of bat may also use street lights as dining areas, though this can make them vulnerable to being preyed on by owls or even daytime birds of prey.

working on ways to reduce light pollution, saving energy and money while helping to prevent these millions of needless avian deaths.

Other ways that too much light affects wildlife occur at a much smaller scale: light pollution around waterways can prevent tiny zooplankton from feeding on algae, leading to algal blooms that choke the life out of the water below. Light pollution is even affecting animals in areas we don't associate with cities, such as the beaches of Barbados.

Turtles have been coming to Barbados to lay their eggs for thousands of years, but now light from the holiday resorts is throwing the hatchlings off track. On emerging, they need to head to the water as fast as they can to avoid being picked off by predators such as gulls. They have evolved to head to the brightest horizon – the water reflecting the light of the moon. But today the brightest glow comes from the hotels and restaurants, which takes the hatchlings in exactly the wrong direction. Red crabs then gather under the streetlights, picking them off before they ever reach the sea. For these turtles, which have seen urbanization take over their once-pristine natural nurseries, the future would be bleak were it not for the conservationists who are helping the hatchlings make it to the sea.

## Big cats and fast food

▲ **Urban young.** Leopard cubs drink from a well in Mumbai's Sanjay Gandhi National Park. Their mother is still catching prey for them, which is likely to include feral animals living off the city's rubbish.

The light coming from our cities at night has increased dramatically in the past few decades, as more and more areas of the developing world become increasingly urbanized. Though this is bad news for most animals, some nocturnal predators are taking stealth to a new level to cash in on the fast food on our city streets.

The people thronging the bustling streets of the Indian city of Mumbai – the most populated city in India – are used to the packs of feral dogs that share their urban home. But there is a far more potentially dangerous creature lurking in the dark: the leopard.

▶ (next page) **Suburban prowl.** A Mumbai leopard on its regular early-evening patrol walks unnoticed by the occupants of the apartments overlooking its path. Their building has encroached on the park where the leopards live, but the leopards have adapted by hunting the dogs, goats, pigs and other animals that accompany humans.

Leopards are lone hunters, pursuing their victims with care and stealth, using the cover of night and often attacking from just a metre or two away. When hunting in the city, a leopard will stay in the shadows, stalking unwary prey. Feral dogs are 25 per cent of the menu (there are a lot in Mumbai), and goats and piglets are taken too, though if a parent pig spots the leopard and sounds the alarm, the game is usually up, as pigs are willing to fight a predator. Urban leopards do very occasionally take humans, usually those living on the city-park perimeter, sleeping on the ground or relieving themselves outside. Generally, though, the noise and crowds protect most city-dwellers. Come daybreak, Mumbai's leopards retreat to the national park that the city has expanded into, returning only at nightfall.

## The leopards of Mumbai

The moment darkness falls, the leopards of Mumbai go out on patrol. Their heartland is the great central park of the city, but as the people have encroached more and more into the park, so leopards have learned to make a living in the metropolis. They hunt mainly pigs, goats and stray dogs but are opportunists and will take chickens and occasionally even humans should they encounter them sleeping out in the open or paying a call of nature under cover of darkness. But far more people are killed by people than by leopards in Mumbai, and for the most part, the human population is unaware of the leopards on their doorstep.

**1 On the prowl.** This thermal image shows a leopard close to illegally built apartments within Mumbai's city park. It is following a regular path to where it knows feral pigs might be found.

**2 The snatch.** Too late, the sleeping sow hears the squeal of one of her piglets, wakes up and chases the leopard in a vain attempt to make it drop the baby.

**3 Safe retreat.** A wall provides the urban equivalent of a tree branch, taking the leopard out of reach of the angry sow below and giving it a moment to suffocate its prey.

**4 Piglet snack.** The leopard takes its prey away to eat in peace. Over several nights, the sow lost all her piglets to leopards. It may be that leopards are using noise from parties in the slum district to help stalk their prey.

## The green cityside

▶ **Sky-high treescape.** A pair of residential apartment blocks in the newly revitalized Porta Nuova district of Milan, Italy, shows how a concrete city can be transformed, upwards. The trees reduce pollution, cool the air in summer and provide the residents with the emotional benefit of seeing greenery and having some contact with nature.

All over the world, cities are expanding to cover more and more of the countryside. And though the pressure to build within the urban areas is strong, there is also a need for green spaces where city-dwellers can relax and even reminisce about the rural homes they have left behind. Many of these green spaces – parks, gardens and other urban oases – are fiercely protected by the people who use them. One result is that these have become havens for wildlife. And now, even newer ways are being explored to make the city landscape better for people and wildlife, including green roofs, in which turf and plants are placed on the tops of buildings. Green roofs are increasingly valued both for their environmental benefits such as water retention and energy conservation, and because they attract wildlife.

In the heart of the north Italian city of Milan, conservationists and architects have created an even more imaginative way of squeezing more green space into the ever-crowded urban landscape by building a 'vertical forest' up the side of a tower block: 700 trees, equivalent to 2 hectares (4.9 acres) of woodland.

Thousands of miles to the east, the city-state of Singapore is even more crowded, with 6 million people squeezed into an area the size of the Isle of Wight. It's hard to get out of the city, so people have invited wildlife in. Everywhere there are green walls and trees sprouting out of buildings, the result of a city subsidy that encourages the creation of a new and bigger green space for the loss of any existing one. Among the corporate giants of Southeast Asia, this has started to set a trend, with the makers of new buildings fighting for the accolade of the most verdant structure.

The greening of Singapore is on such a large scale that it now has more variety of wildlife than any city in the world. Singapore is also one of only three cities that are classified as biodiversity hotspots.

Milan and Singapore have shown us that we can incorporate a diverse ecosystem within our cities, if we have the will to do so. But this is still the exception, not the rule. More people live in cities today than in rural environments, and this proportion is only set to increase. We have the power to shape our cities to benefit both us and wildlife, should we choose to do so. We are in control and can use our powers of ingenuity and imagination to welcome the wildlife in and reap the benefits that living alongside nature has on our physical, emotional and spiritual health.

# 7
# TALES

**TEN YEARS AFTER THE ORIGINAL** *Planet Earth* series hit our screens, the challenge for *Planet Earth II* was an awesome one. How could it possibly outdo, or even match, the original? By using a new approach, a new perspective and taking advantage of new technology – that was the view of series producer Tom Hugh-Jones, who also worked on the first series.

From the start, the aim was to give viewers the sensation of being actually in the animals' environment, which meant filming, editing and writing to create an immersive experience. Rather than watching animals through a long lens of a camera, the aim was to make the most of new equipment and cinematic techniques to take the viewer into the landscape and among the animals. So though not every story in the series is new, the way in which it has been approached is.

Today, image-stabilized cameras can be used in many more ways than on moving vehicles. They have become more compact, lightweight and reliable, which means they can be hand-held and used on drones, giving 4k images (twice the definition of the usual high-definition images) while on the move.

On the subantarctic island of Zavodovski, for example, as on many other shoots, the high-definition camera and its wide-angle lens was carried on a special stabilizing rig that allowed it to make steady images whichever way it was tilted. So it could sweep us at low level through the vast chinstrap penguin colony – the biggest penguin colony on Earth. 'It's an amazing spectacle,' says Tom. 'One that few people have ever seen, let alone experienced in this way.'

(previous page) **Catching the catcher.** Having followed a red fox in Yellowstone National Park, Wyoming, long enough for her to be habituated to his presence, cameraman John Shier films her catching voles and mice, which she does with a pounce, head first into the snow.

◀ **Night vision.** As dusk falls in Mumbai, Gordon Buchanan sets up his camera – an ex-military piece of kit that records in infrared the heat signature of any mammal. He will then retire again to his hide to await the appearance of a leopard. His watch lasted 28 nights.

Another unforgettable sequence in the Islands episode is that of the baby marine iguanas being pursued by hordes of racer snakes. This, too, was shot using a wide-angle lens on a camera mounted on a stabilizing hand-held rig. The story – cut together in the way a movie sequence would be – shows not only fascinating behaviour but also the quirky reptiles down at their level.

Of course, you can't script and direct wild animals, and traditional long-lens filming is needed to record much of the behaviour. But the cameras allow low-light recording that wasn't possible ten years ago, and lenses have also changed. Rather than giving the impression of looking down a microscope, today's macro and scope lenses give a much wider depth of view, and the kit has shrunk, allowing much closer access to little animals – ants, locusts or even harvest mice living in a forest of grass. And because of the increased light sensitivity of the cameras, you can get close to animals, such as the little glass frogs in the Jungles episode, without the need for powerful frog-frying lights.

Where sweeping vistas were needed to introduce us to the landscapes, the Namib Desert, for example, and where animals would not be scared by flying cameras, drones were used. The latest ones were field-ready for the team: safe to fly and reliable, with a long battery life and stable in wind.

New technology also helped the production team on the Mountains episode to film the snow leopard. One of the few truly iconic mountain animals, it featured in the original *Planet Earth* series, with a hunt sequence so memorable that the team had a lot to live up to. But with the help of local experts – a key asset for any wildlife film-maker – and the use of the latest remote cameras, they achieved their aim of filming new behaviour.

The synchrony between style and content reaches its zenith here. The story is intimate both in how it looks but also in what it reveals, with astonishing wide-angle close-ups of snow leopards as they brush past the cameras, and intimate details of their courtship. This was possible because of remote-camera technology that can now film high-quality moving images, but also because enough cameras were carefully positioned in the field to allow a full sequence to be shot and edited together.

► **Reflecting.** On the island of Escudo de Veraguas, a mother pygmy sloth peers at the camera-rig. Completely unfazed by the cameraman, who has crawled towards her through the mangroves, she is looking at her reflection in the wide-angle lens, as is her baby. With no experience of predators, the island sloths have little fear of humans.

▲ **Trial flight.** The drone team trial a new development – a camera mounted on a 360-degree 'virtual reality' rig on top of the drone so it can look up as well as down. Guy Alexander drives the drone and Ewan Donnachie moves the camera and focuses. The top-mount rig was used to film some of the jungle canopy and canyon scenes.

◄ **Double-checking.** Camera assistant Louis Labrom checks the shots that have just been taken of macaques raiding the food market. One of the macaques, ever inquisitive, closely observes.

In many cases, the subjects were animals that were relatively unafraid of humans. Island species in particular allowed an up-close-and-personal technique. For the Cities episode, the stories were about animals that manage to live alongside people, whether macaques in India or peregrines in New York, but the filming of leopards in Mumbai was achieved not by being at close quarters – impossible at night – but by using the latest military thermal cameras, which can now shoot high-resolution footage good enough for storytelling.

In the Cities episode especially, says Tom, 'We become aware not only of the way some animals have adapted but of the extraordinary tolerance that some human city inhabitants show towards the wild creatures that live alongside them.' It also reveals our impact on the lives of animals.

On the Caribbean island of Barbados (as on beaches elsewhere), hatchling turtles are so confused by the lights of beach developments that they head away from the sea rather than towards it, and the shoots at night mostly recorded baby turtles dying in street gutters.

*Planet Earth II* features some incredible spectacles – the caribou migration, for example. Tom fears that, in another ten years, these spectacles may no longer happen in the same way, because of developments blocking migration paths or encroaching into refuges or because of even bigger changes resulting from the disruption of the climate. 'People need to understand that the world's wildlife is in big trouble,' he says. During the making of this series, the team faced many planning challenges caused by unseasonable weather, including shoots that had to be cancelled or postponed (three seasons in a row in the case of the Ecuador hummingbird shoot). 'No longer are natural events as reliable, as they were when the original series was filmed,' says Tom. 'It's now almost the norm for the climate to be unusual.'

Tom would like *Planet Earth II* to inspire a new generation of viewers – and there will be tens of millions of them – to marvel at the world's wildlife spectacles, and he hopes that their emotional response to what they see will drive further action to protect nature and secure its future. There is also the so-called *Planet Earth* effect. After the first series was shown, many of the places featured were visited by viewers who wanted to see the wildlife spectacles for themselves. Responsible wildlife tourism can help safeguard at least some of the special places and creatures the series has filmed, such as the indri, the world's largest lemur, which travels through the trees on its elongated arms like Spider-Man. The hope is that an emotional response to what they see in this series may prompt people to support the work of the visionary conservationists trying to replant the Madagascar rainforest or to persuade communities or governments to protect the last populations of precious species and the habitats that are their homes.

▶ **Peregrine angle.** John Aitchison waits for a pigeon's-eye view of a peregrine falcon when it finally decides to fly from its nest, while expert Matt Wilson keeps watch and producer Fredi Devas tries out his camera. Behind, New Yorkers pause to watch the sidewalk tactics.

▶ (next page) **Camera test.** In the community-run Analamazaotra forest conservation project in eastern Madagascar, an inquisitive indri examines the camera rig held by researcher Emma Brennand. The lemur, named David after Sir David Attenborough, was used to people, but Emma had not expected such a close encounter or for him to be so interested in camera gear.

RED PRO
5.0

## Catching the ghost cat

As producer Justin Anderson explains, if you make a show about mountains, you just have to feature snow leopards. He was determined not only to film these iconic creatures, but also to reveal new aspects of their lives.

But the snow leopard is legendary for being one of the most difficult animals to film. On *Planet Earth*, veteran wildlife cameraman Doug Allan spent weeks sitting in a cave waiting for this elusive animal to appear – which it never did. Later his colleague Mark Smith did manage to obtain a memorable sequence of a single individual, but only after many more weeks of effort and frustration.

Preparation is the key to filming any wild animal, but with snow leopards, it is absolutely essential. Fortunately the team had excellent contacts throughout the cat's range, who could provide up-to-date information. First came bad news: at the site in Pakistan where the *Planet Earth* sequence was eventually filmed, snow leopards were no longer present. Having checked out sites in Mongolia, the team finally struck gold: tourists visiting Hemis National Park in the northern Indian region of Ladakh were returning with excellent photos and videos of snow leopards, some even taken on smartphones.

The following spring Justin flew out to Ladakh with wildlife cameramen Mateo Willis and John Shier. They also took Duncan Parker to set up the many remote cameras essential for capturing footage of the leopards, and Sue Gibson, who would shoot the team making the film. They then enlisted the help of Khenrab Phuntsog, who works in the park as a wildlife guard. He and his colleagues were to prove invaluable.

The trip began promisingly, with nine animals seen in the first five days. One reason for the team's success at finding them was that, instead of staying put in a hide and waiting for a snow leopard to walk past, they had spotters out at dawn every morning, who radioed the cameramen so they could be in the right place – not an easy task given the very steep and tricky terrain.

Gradually a story emerged: each of the snow leopards they located would follow a specific trail, and some of the trails crossed. The cats rarely

▶ **Star number one.** The female star of the Mountains episode walks her regular route, triggers a light beam and is filmed from a carefully positioned camera. Her route was located by the expert trackers and the camera placed precisely to film her facing it.

▲ **Scent talk.** A male snow leopard sniffs the spray-height pee-message left by another male on a rock beside a regular snow leopard path.

◄ (top) **Big-cat stand-in.** Camera assistant Duncan Parker acts out being a snow leopard while cameraman Mateo Willis adjusts the camera focus.

◄ (bottom) **Cold light.** Khenrab Phuntsog, one of the team's vital local experts, sets up a camera-trap with cameraman John Shier. All the locations had white snow and pale rocks, which made choosing the right camera settings for the lighting a constant problem.

met but they did leave signals and signs for one another in the form of 'pee-mails', urine sprayed on rocks that convey information such as the age and sex of the animal, and whether or not a cub is accompanying it.

Altitude is always an issue when filming snow leopards, and this site was no exception: the base camp was at 3500 metres (roughly 11,500 feet), while the team occasionally had to venture up to about 5000 metres (16,400 feet), where altitude sickness can be a real problem.

Another problem, says Justin, is that 'although we'd see the cats, they would be inactive during the day, mainly sleeping; and then just as dusk fell and it got too dark to film, they would wake up and start doing really interesting things.'

But the team did manage to follow a mother and cub for several days in a row, and it was Duncan's camera traps that really made the difference. More than 20 were placed at strategic points around the snow leopards' ranges, and together these helped build up a picture of the complex social life of these seemingly unsociable animals. The remote cameras used for camera traps are both higher quality and far more reliable than they used to be, and produced footage that, says Justin, showed snow leopards in a very different light. 'Though the animals don't encounter each other very often, they still have a social network: they know what is going on and which other animals are around.'

Though the female they were following had a cub, it was almost weaned, so when she came across a new male, she was ready to breed again. At first the male kept his distance, following the scent marks she left on her regular circuit. But gradually he got closer and closer. 'When she was ready for him, she went to a high point and started to sing – a truly incredible sound, which echoed around the mountains. Amazingly, we managed to capture this with one of the camera traps. The male had been led on a merry dance until now, but finally this was his chance to mate with the female.'

It was now that the cub made herself scarce (male snow leopards, like other big cats, will sometimes attack and kill any offspring not their own). But then a second male arrived and intercepted the cub. The mother and her mate also arrived, which gave the team the chance for a unique sequence.

The female was now facing a major dilemma: whether to protect her cub or continue to mate. It all ended up with a fight, in which the cub ran off and the female got wounded. 'We thought that was the end of the story,' says Justin, 'but a few months later, we located her again, and she was OK. The cub was still following her, but now about ten minutes behind, rather than close by.'

The team did return the following spring, but a lack of snow meant that the cats were at far higher altitudes than normal and so considerably harder to film. But using the camera traps again, they were able to complete the story.

It was a gamble trying to film behaviour, but it paid off. 'As the plane took off on my final visit,' says Justin, 'and I looked out of the window at the Himalayas stretched beneath me, I realized just how privileged we'd been to spend time with these beautiful wild creatures.'

► **Keeping in touch.** The female star of the film is greeted by her nearly independent daughter. Keeping track of her movements high up in the mountains of India's Hemis National Park, Ladakh, were three spotters, linked by radios to the camera operators, but it was the camera-traps that eventually caught the most intimate action.

## A cast of billions

'Heading north once more into even more remote Madagascar to try to find locust swarms. Who knew that finding several billion locusts would be this difficult?' So wrote producer Ed Charles on his blog, while desperately trying to catch up with one of the most destructive creatures on the planet.

They may gather in their billions, but locusts are unpredictable, and even a huge swarm can be hard to track. So Ed and his colleagues enlisted the help of the United Nations Food and Agriculture Organization (FAO), which also needed to pinpoint the location of these marauding insects – before they headed off to wreak crop-destruction elsewhere.

Madagascar is one of the poorest and least-developed countries in the world, with roads that are often little more than dirt tracks. After any

▲ **Grounded.** The helicopter that landed the crew ahead of the locust swarm. Once the locusts arrived, it was too dangerous to fly, though when the swarm moved on, the crew did film from above, out of the open door.

sudden downpours – such as the rains that bring the locust swarms – they become impassable. Time and again, the BBC crew found themselves tantalizingly close to the locusts, yet unable to reach them. Things weren't helped by the fact that most of the team went down with dysentery; and in this remote area, there was a very real danger from armed bandits. In the end, the only way to reach the swarm was to charter a helicopter.

The spectacle turned out to be one of the most incredible they had witnessed. 'The thing that struck me most was the sound,' says Ed. 'At close range I could hear the almost metallic clatter of individual wing-beats as they brushed past each other, but beyond that was something colossal – a deep roaring noise almost on the edge of hearing, created by billions of these wings beating in unison, displacing vast quantities of air as they rushed over the land.'

Though he was well aware of the locusts' potential for destruction, he was struck by how incredibly beautiful the enveloping swarm was. 'It was almost hypnotic – the gentle roaring sound combined with the abstract patterns – an incredibly serene experience.' He and the team stood in awe, 'almost forgetting that we should get on and film something'. Cameraman Rob Drewett described the swarm as 'one shimmering, silver organism'.

To capture the full spectacle, cameraman Justin Maguire took to the air in the helicopter. From a distance, the locusts 'seemed like smoke from a bushfire'. As he got closer, the challenges of capturing such an all-encompassing event became clear. 'Once in the swarm, my adrenaline rushed as the cloud rapidly moved and changed, and I tried to get the best shots, not knowing which angle would best portray this spectacle.'

This was one of the biggest locust migrations ever recorded, with swarms more than a mile wide infesting about half the island – an area more than twice the size of the United Kingdom. A locust may weigh just 2–3 grams (0.1 ounce), but it can eat three times its own body weight every single day. That means that an average swarm – a billion locusts – can consume as much as 9000 tonnes of food, day after day.

And this puts the livelihoods of 13 million people already living in extreme poverty at great risk. As Ed notes, 'It's hard to imagine the devastation that a plague of this scale can leave in its wake. I've seen a field of corn that took months to grow stripped in a single day, destroying the livelihood of one family.'

The team was able to offer some help, by reporting their sightings to the FAO team, so that they could monitor the swarms and work out where the locusts were heading next. But it was still hard to feel elated by their success, knowing that the swarms were laying waste to the land. Since filming finished, there has, however, been some good news for the farmers of Madagascar. After three years, the plague of locusts now appears to be in remission, meaning that the eradication efforts of the FAO have – for the moment – succeeded in halting the terrible destruction wrought by these incredible insects.

▶ **Joining the swarm.** Cameraman Rob Drewett runs around in the swarm, filming at locust flight-level with a wide-angle lens. The rig was just light enough to hand-hold, stabilizing his movement and buffering any bounces so Rob could move with the insects in a way that was previously only possible by using a huge crane-like jib arm.

## Penguin paradise, human hell

For almost six decades, the BBC Natural History Unit has pushed the boundaries to film wildlife in some of the toughest environments on Earth. But few stories have presented quite as many challenges as filming penguins on Zavodovski, an active volcano and one of the remotest of the subantarctic islands. Planning a shoot on such a challenging location took a year.

Home to the largest breeding colony of penguins anywhere in the world, Zavodovski Island promised a truly spectacular sequence – if only the team could get safely on and off this storm-swept speck of rock. To have any chance of doing so, they enlisted the help of Jérôme Poncet, who knows the seas and islands around the Antarctic Peninsula better than anyone.

**Weathering it.** An eye-level view of the colony. While one partner sits tight on the chicks, the other goes fishing, returning with a full crop to regurgitate to the chicks. Every penguin on nest duty has its back to the wind, receiving an icing of snow when a blizzard blows in. With no other landmass to block the winds, storm after storm blows across the island, leaving just one in four days fine enough for filming.

The plan was to film for two weeks in late spring, in January, when the chinstrap penguins had chicks.

All Jérôme's 40-plus years of experience were needed just to get the team onto Zavodovski in the first place, as producer Elizabeth White explains: 'Jérôme and his yacht the *Golden Fleece* were absolutely crucial to the success of our mission. After seven days crossing from the Falklands over the roughest ocean on Earth, we finally caught our first glimpse of the island – quite a surreal moment, given that we had spent a whole year putting the expedition together.'

Getting *to* the island was one thing – actually managing to make landfall quite another. Since explorers first discovered Zavodovski on Christmas Eve 1819, few others have visited this grey lump of volcanic rock, one of the

South Sandwich Islands, 350km (217 miles) southeast of South Georgia. Now the BBC team not only had to land, but also get all their filming gear and supplies safely on land as well.

Jérôme is the one person in the world who knows the island well, and knew there was one potential landing site: a wave-splashed rock face. With unusually calm sea conditions, they decided to take their chance – all the while knowing that if anyone slipped and fell, the expedition would have to be abandoned. Getting all the gear onto the island, using a system of ropes and pulleys, took a whole day – and the process was not helped by the presence of so many penguins. 'The hardest thing for us', as cameraman Max Hug Williams points out, 'was finding a pathway through them – there were penguins absolutely everywhere you looked.'

▲ **Mother yacht.** As the *Golden Fleece* drops anchor off Zavodovski, Jérôme (right) and crew-member Yoann Gourdet check out a possible landing area for the Zodiac boat.

▶ **Only way in.** On a relatively calm day, at the only possible landing point, the crew haul a ton of provisions and filming gear up the volcanic rocks and along the side of the cliff to where they could reach the plateau without disturbing nesting penguins.

▲ **Sunny view.** Parents on duty with their chicks on a rare sunny day, revealing how their nests are just beyond reach of both pecking and excreting. Behind is the steaming top of the volcano.

◄ **The long view down.** Cameramen Max Hug Williams and Pete McCowen film the penguins riding the huge waves below as they try to land on the cliff. A few macaroni penguins – more inquisitive than chinstraps – pause to watch the filming.

The team did eventually manage to pitch their camp on a relatively penguin-free spot – though there were still plenty of nosy neighbours coming to take a look at these unexpected intruders. Having never seen a human before, the curious penguins showed no fear.

At last, the team could get down to filming this spectacular penguin colony. Max Hug Williams was blown away by what he saw: 'As soon as you walk over the ridge and see the chinstrap colony, you get a tingly feeling – I've just never seen so many animals together in one spot. This must be the most photogenic place in the whole world – it's truly mind-blowing. Now we had to do it justice.' The team's other cameraman, Pete McCowen, compared the sight to Glastonbury Festival – but with, instead of people, penguins – more than 1.3 million of them.

All was going well with filming until there was a sudden change in the weather, with storms bringing high winds, snow and heavy waves. So the filming had to be put on hold.

When the storms died away, and the weather warmed a little, another problem presented itself. Though the melting of the snow meant that they could start filming again, the bad news was that they discovered why their campsite was free from nesting penguins – it was right in the path of the melting snow-water mixed with penguin excrement.

Each day presented a new and more difficult challenge. An even heavier storm forced the team back to the safety of their tents for two days. At least they were relatively warm and comfortable, but the penguins had no choice but to go out to sea to catch fish for their hungry young. This resulted in a heart-breaking spectacle, as the penguins struggled to survive the massive waves.

'The beach was a scene of death and destruction – absolute carnage,' says producer Elizabeth White. 'They were trying so hard to get up the cliffs, covered in blood and with broken legs.' Cameraman Pete McCowen witnessed the penguins being battered against massive boulders and being catapulted 15 metres (50 feet) into the air on the huge waves.

With such a powerful sequence in the bag, the team were ready to go home, but now they faced their biggest problem of all. The storms had left a massive swell, which meant it was difficult to launch the Zodiac boat to pick them up. But with only a few days' food and water remaining, and the weather forecast predicting a worse storm, bringing an even heavier swell, they had no choice – they simply had to get off the island before the short window of opportunity closed. So using all his experience, Jérôme manoeuvred his inflatable Zodiac boat close inshore, timing his approaches with the huge waves and the wash. Now began the tricky task, as the swell grew, of ferrying all the kit on board with ropes, followed by the team itself.

After two hours, thanks to Jérôme's extraordinary skill, they were all back on the *Golden Fleece*, with one of the most incredible of island spectacles safely in the can.

▶ **Basic base camp.** The pitch – the only penguin-free site. It had spray from storm waves but shelter from the katabatic winds that would have shredded the tents. When it rained, it became obvious why penguins didn't nest there: runoff turned it into a sewage bog.

## The antelope apocalypse

Wildlife film-makers are used to having to deal with the unexpected: transport problems, bad weather and political instability often mean that shoots have to be cancelled or curtailed. But when producer Chadden Hunter went to film vast herds of saiga antelopes on the steppes of Kazakhstan, he could never have imagined what he would find.

Scientists have long been aware that saiga populations go through a cycle of boom and bust. And since the breakup of the USSR, a market for horns in China and meat in Russia and a breakdown in law enforcement

▲ **Camp saiga.** The team and the scientists eat supper, having set up base camp just a couple of kilometres' walk from the saigas' birthing ground. The ex-Soviet military truck not only served as their overland transport as they searched the steppes for the saiga but also as their kitchen and film office.

caused a 95 per cent decline. There used to be at least several million of these unusual antelopes roaming the flat, featureless grasslands of central Asia, but by 2000, there were only about 26,000 left in the wild (three populations in Kazakhstan and one in Russia, with a tiny population of a separate subspecies in Mongolia).

But the saiga population had bounced back, rising to between 200,000 and 300,000, the majority in Kazakhstan. So by the time Chadden and the team arrived, with Kazakhstani and German researchers, they were confident that there would be large numbers. The team had come to film the annual calving – a brief window of opportunity in the spring, during which every female gives birth in the same few days, using safety in numbers to maximize each calf's chances of survival.

Chadden was accompanied by Martyn Colbeck, one of the world's most skilled and experienced wildlife cameramen. Martyn knows the saiga better than almost anyone, having first filmed them in 1989 for the ground-breaking series *Trials of Life* and then again in 1991 for *Realms of the Russian Bear*. He wanted to go back to film them for *Planet Earth* in 2005, but by then the saiga population was at its lowest, and their future was far from secure. So when Chadden asked him to film saigas for the series, Martyn jumped at the opportunity. Not that this vast, open, big-sky landscape is an easy place to work.

The team's first job was to find the herd of females and find where they were going to give birth. That they were able to do this was because of the radio-collars the scientists had fitted on some of the saigas.

After giving birth, female saigas leave their twin calves hidden in the grass for a day or two until they are strong enough to walk. To film the calves, the team would be out on the steppes at 3am, digging a hole for Martyn and all his gear to spend the day in, so that he wouldn't disturb the skittish animals. But even before they began filming, Martyn sensed that something was not right.

'On the way to the calving grounds we spotted the occasional saiga carcass way off in the distance, shimmering in the heat haze. One would expect the odd saiga to die of old age or disease, but as the journey progressed we saw a few more carcasses, and our scientist guides started to look concerned. I tried to recall if I had seen any carcasses on my previous visits, but didn't think I had.' By the third day, there were literally hundreds of carcasses.

The team continued filming, and from his subterranean hide Martyn managed to get intimate footage of the calves suckling. A fine sequence was almost in the can, but still the animals were dying. Martyn recalls: 'Day after day the number of carcasses increased. What was really strange was that there appeared to be very little external evidence that these animals were sick. The carcasses showed evidence of diarrhoea and some

▶ **Hide Martyn.** Cameraman Martyn Colbeck about to be released from 15 hours in a foetal position in his small, hot hide. It had been a good first day for him, having filmed calves being suckled early in the morning. But behind the hide were some of the first dead females.

bleeding from the nose, but otherwise they looked perfectly healthy.' As if this wasn't enough, a huge storm then swept across the whole area, and the team had to retreat to camp. After it had finally passed, one of the scientists went out to check on the status of the 100,000-strong herd – the largest known herd in the world. When he returned to camp he simply said, 'It's over. They've all died.'

When the team returned to the calving grounds the next day, Chadden could hardly believe his eyes. 'It was like Armageddon. There were tens of thousands of saiga bodies lying all the way to the horizon.'

During his long career as a wildlife film-maker, Martyn had never seen anything quite like this. 'It was very sad and disturbing to witness. Calves were even attempting to suckle from their dead mothers. The situation became worse as the steppes became littered with orphaned calves wandering around in search of their mothers. It was heart-breaking. In all my years of filming I had never witnessed anything remotely like it. It was totally baffling – hard to imagine what could cause such a pandemic.'

It was not just the spectacle but a single moving event that Martyn found the most distressing. 'The most poignant moment for me was when an orphaned calf came right up to my buried hide in search if its mother and stood inches from me bleating. There was nothing I could do for it.'

Soon afterwards, the military arrived, and became understandably concerned that what they assumed was a BBC news crew was filming this massive natural disaster. Despite the team's protests that they were only interested in the healthy calves and females, the military demanded that the team stop filming and hand over all their precious footage. Chadden began duplicating Martyn's shots so that they would have a back-up. It was a risky strategy, but it worked.

Before they left, they witnessed an extraordinary clean-up operation, in which the military buried tens of thousands of animals in vast pits. 'It was like something out of *The X-Files*,' says Chadden. 'They had bulldozers digging mass graves the size of a house, huge dump trucks and teams of guys in white biohazard suits dragging the saiga into piles so they could be buried. The next morning there was no evidence the saiga had ever been there.'

◀ **Before the births ... and deaths.** One of the German scientists takes a picture of the herd grazing on the flush of grass. It was day one of filming, and the normally highly nervous saigas had drifted towards camp and were within 200 metres. Hopes were high.

◄ **Casualty.** The first dead saiga to be found – a male. Its translucent horns, so prized by poachers, were intact, and there were no signs of who or what had killed it.

► **Motherless.** A starving calf approaches Martyn's hide and microphone, hoping to find something to suckle – part of the heart-breaking aftermath of the overnight deaths of thousands of female saigas.

When they finally returned home, the story of this extraordinary event began to reach the western news media. Eventually Chadden discovered that, in little over a week, 70 per cent of the world population of saigas – at least 200,000 – had died, putting this unique and already endangered species at some risk of extinction.

One theory is that the wet, warm spring weather, which triggered the grass growth, changed a normally harmless bacterium carried by saigas into a lethal one. Certainly two discrete subpopulations many kilometres apart were affected simultaneously, which points to an environmental trigger and not an infectious agent.

Yet despite the cataclysmic events they witnessed, Chadden remains cautiously positive about the future of this incredible animal, partly because of the nature of the saiga's biology – it has recovered from similar though smaller die-offs in the past – and partly because of the efforts that the Kazakhstan authorities are making to protect the remaining animals from poaching. Says Martyn, 'Saiga populations have fluctuated wildly in recent years and they clearly have the capacity to rebound quickly, given that the females can reach sexual maturity at less than a year old and most can produce twins. So I feel optimistic that once again the saiga will recover and continue to roam the steppes of central Asia.'

## Filming the dark-water dolphins

One of the most extraordinary stories the Jungles team came across was the discovery of a new species of mammal in a remote area of the Amazon. New species of insects and other small animals are regularly found in tropical forests, but a newly discovered large mammal seemed an unmissible opportunity to illustrate the diversity and surprises that make the rainforest so special.

Producer Emma Napper and cameraman Tom Crowley ventured into a flooded forest in the heart of the Amazon Basin, one of the toughest places in the world to film. Their aim was to film the Araguaia river dolphin, a species only known to science since 2014. But they soon found out that this wasn't going to be easy. These river dolphins are shy and spend most of the time under water, only surfacing for a few moments before vanishing into the murk. And, of course, they weren't in open water but swimming between trees in the depth of the flooded forest – the chances of success were small.

▲ **Navigating the flooded forest.** With a stabilizing rig on a jib arm fixed to a canoe, the team searches for the elusive Araguaia river dolphin.

The first problem was finding the dolphins. The scientists studying river otters in the area had seen the dolphins while travelling through the flooded forest. But they weren't monitoring them daily. As Tom recalls, 'It was one of those situations where everyone says, "You should have been here yesterday!" Wherever we went, the dolphins had moved on.' So they enlisted the help of a local guide Juarez, from the Instituto Araguaia, who proved remarkably good at understanding the animals' behaviour. As Emma says, 'Always get the help of the people who have lived there since they were kids – they are the ones that have built up a picture of where they go at what time of day.'

After several days, they started getting glimpses of the dolphins among the trees. But filming them was almost impossible. The water in the Araguaia is incredibly murky, so underwater cameras were useless. Also filming had to be done from an inherently unstable canoe. And in the river were huge catfish that, at the surface, looked uncannily like dolphins. As Emma realized, there was a good reason little is known about these animals: 'They are like ghosts. The only way to spot one was to look for little signs in the water and then hope you get the camera on the right patch of water a couple of seconds before a dolphin comes up to breathe.' Tom agrees: 'The problem is that you only know they are there if you hear them, but often if you hear them, that means you've missed them, as they have already gone past.'

The flooded forest is flooded for a reason. The crew also had to endure a massive storm that appeared out of nowhere. On another occasion, torrential rain was followed by massive hailstones falling onto their exposed canoe – the first time the locals had ever seen hail. As the storm passed, Tom recalls, 'There was a beautiful orange sun coming through the veil of rain – a once-in-a lifetime moment. But, of course, the camera was safely packed away.'

Despite the frustrations, they were starting to learn about the dolphins' movements. But to keep up with them, they needed to build a special rig, a gyroscopically balanced tripod that could be used even on the unstable canoe and enabled them to shoot with a long-focal-length zoom lens. They also had a drone, and this was to reveal something new. The dolphins were neither solitary nor lived in groups of one or two, as had been supposed. They live in loose pods containing several individuals. 'From the drone you could see the shapes of dolphins beneath the surface. Suddenly we realized

that there weren't one or two next to the boat but seven or eight. It was a completely different perspective. They surrounded the boat, and yet from the surface, you couldn't see a thing.' Tom would have liked to have gone into the water to try to film them, but the presence of caiman, stingrays and giant fish capable of taking your leg off in one bite meant that this was not a viable option. They did drag a small camera behind their boat, but the lack of visibility – less than a metre – meant that they only got a couple of usable shots.

The dolphins might have been hard to find, but plenty of other animals were all too easy to see. The crew stayed on a small spit of land in a little hut on stilts. 'Inside with us was every animal that wants to keep dry: spiders, scorpions and big rats. The hut was simply crawling with life.'

By the end of the shoot, Tom and Emma had finally achieved their near-impossible aim: a sequence to show a creature new to science that few people have seen. As Emma says, 'It's definitely the hardest animal I've ever filmed – and we knew all along that if we didn't get enough shots we didn't have a sequence, and all that time and effort would have been wasted. So, yes, I feel pretty good about what we achieved.'

▲ **Waterway wide-shot.** While producer Emma Napper manipulates the jib arm, Tom Crowley watches the shots and angles and focuses the camera.

▶ **Final sighting.** The head of a river dolphin surfaces for two seconds for a quiet breath. This was as near as they ever got to one in the canoe.

◀ **Night-shift bone-cruncher.** A hyena pauses beside the police station to sniff and listen before going through the narrow entrance to the meat-market square. At night, when the markets are empty of people, Harar's alleyways echo to the whoops of hyenas descending on the square.

## Hyena city

If asked to pick one mammal to avoid at all costs, most people would list the spotted hyena pretty near the top. As the second-largest land predator in Africa, hyenas are feared and loathed. Unlike lions, they are seen as vicious, crafty opportunists, always on the lookout for an easy meal. So when producer Fredi Devas heard about the extraordinary goings-on in the ancient Ethiopian city of Harar, he was intrigued: instead of keeping the hyenas out of the city, the people of Harar welcome them in – and then feed them.

It seems that the people who built the city walls more than 400 years ago knew the advantage of having a clean-up squad in the city. They left hyena-sized entrances for them so they could enter the city to eat leftover bones, and in times of war, they may have even disposed of bodies before they went bad and spread disease.

The logistics of filming in any city can be tricky – security is always a problem, along with the safety of the team and the welfare of the people and animals being filmed. Filming in Harar presented its own special difficulties. The streets are very narrow, with winding alleys and stone staircases – and crowds. And as wildlife cameraman John Aitchison remembers, 'For the first time in my life I had the spine-tingling feeling that a large animal was following me in the dark, and with such soft paws that I couldn't hear it move. This must have been something our ancestors knew well, and dreaded.'

Using a special low-light camera, Fredi, John and their fixer Ramadan headed out each night to find and film the hyenas in little groups as they fed on bones and scraps in the meat market in the heart of the old city. For John, who had filmed hyenas on the African savannah, this entailed a rapid reassessment of his attitude towards these usually wary animals. 'During one of our first nights, I was kneeling in the darkness when I felt a bump on my hand, which was gripping the tripod handle. I expected to find a person, but instead there was a hyena looking me straight in the eye. After a pause, it walked around the tripod and on towards its usual feeding place.'

As always when filming in cities, the team encountered some unexpected problems. Hearing that a BBC film crew was in town, and with local elections imminent, the governor of Harar decided to smarten the place up – which meant that every night decorators converged on the market square to give it a makeover. This caused all sorts of continuity problems, as the filming backdrop kept appearing in a new and different set of colours.

The team's other cameraman, Louis Labrom, did manage to get some remarkable footage of the hyenas gathering and being fed. But this was only half the story, as John explains: 'After the meat market, it was quite a relief to switch to filming the hyenas outside the walls. In town, the hyenas have a truce, and members of different clans pass peacefully, sometimes even greeting each other, but it's a very different matter outside.'

The team found a dozen hyenas patrolling the cobbled streets outside the walls, like a gang looking for trouble, glowing yellow as their spotted coats passed through a strip of lamplight, then fading back into the shadows before entering the home area of another clan. It was now, with only one or two streetlights providing illumination, that the low-light camera came into its own, enabling John to capture remarkable footage of a territory war that involved more than 100 animals. But though the encounter looked violent, the hyenas held back from seriously hurting each other.

For Fredi, the most astonishing moment came when they had finished filming for the night. They were sitting in the kitchen of Yusef, a fifth-generation hyena man. 'Yusef's daughter-in-law was sitting on the step, breastfeeding her baby, who must have only been a few months old. A hyena came into the kitchen to feed, then wandered over and sniffed the baby's head. Yet the mother simply carried on breastfeeding.'

John, too, gained a new insight into the relationship of hyenas with us. 'In the alleyways and market places of Harar, hyenas cross the boundary between their wild world and our urban one. These trusting, and trusted, animals give me hope that something similar might be possible elsewhere – that large predators and tolerant people could learn to trust each other.' Fredi agrees: 'This is the most extraordinary example in the series of a mutually respectful relationship between humans and wild animals.'

▶ **Eye-levelling.** John Aitchison on night duty in the meat-market square, using a low-light camera at an eye-level perspective to film hyenas crunching up the bones. If there were scraps of meat to be had, the ever-present dogs (refuse collectors) would charge the hyenas in fits of bravado. Surrounding the square are the butchers' shops and stalls.

▶ (next page) **Going for first light.** Tom Walker skirts the edge of the Ramon Crater in Israel's Negev Desert, looking for the best spot to film Nubian ibex. They are still asleep on a cliff-ledge below but will soon wake and climb down to drink in a gully. Tom is wearing an 'easy rig arm' to steady the camera in the high wind as he climbs down to his filming position.

I·D·II